HOW TO READ A ROCK

Published by Smithsonian Books
Director: Carolyn Gleason
Senior Editor: Jaime Schwender
Assistant Editor: Julie Huggins

Edited by Natalia Price-Cabrera
Designed by Luke Herriott

This book may be purchased for educational,
business, or sales promotional use.
For information, please write:
Special Markets Department, Smithsonian
Books, P.O. Box 37012, MRC 513,
Washington, DC 20013

Library of Congress Cataloging-in-Publication
Data available upon request.

Printed in China, not at government expense
26 25 24 23 22   1 2 3 4 5

HOW TO READ A ROCK
Our Planet's Hidden Stories

JAN ZALASIEWICZ

Smithsonian
Books

Washington, DC

CONTENTS

5 Rocks as storytellers

6 Human-made rocks

7 Rocks on other planets

INTRODUCTION

Our planet is nearly 13,000 kilometres (8,078 miles) across. It has a thin skin of water, in oceans, lakes and rivers, and an even thinner living skin of soil and vegetation, with an atmosphere above. But Earth is nearly all rock, from which the water, air, soil and life have all emerged, and on which they are utterly dependent. Rocks are the foundation of our lives.

As a child I knew none of this, but I found rocks endlessly fascinating. They seemed like a gateway to worlds beyond even the fevered imaginations of Hollywood scriptwriters. Scrabbling around in stream beds, I could unearth mysterious mineral patterns and the petrified remains of animals and plants that died millions of years ago, which brought with them a host of puzzles and enigmas. I was captivated then and, after a career among rocks, I am still captivated by their infinite variety and richness.

Rocks tell stories of past worlds. They tell of landscapes and seascapes inhabited by dinosaurs and giant marine reptiles, by trilobites and corals – even the endless microbial layers of primeval Earth.

But these stories can stretch much further. Using clues gleaned from the rocks, we can evoke the speed and strength of currents that flowed in vanished rivers and long-dried seas; recreate the forces unleashed within a speeding avalanche or a meteorite strike; follow the path of white-hot magma through Earth; or simply peer several kilometres underground to where, for millions of years, subterranean fluids have trickled past and, with infinite slowness, deposited tiny mineral gardens in the spaces between sand grains. The stories held within rocks have no end. You can discover them and read them for yourself, given a little time, a sense of curiosity and a yen to puzzle through the many clues on those rocky surfaces.

Today, we live in a golden age of rock – far more so than did our Stone Age ancestors. They lived in a world clothed in forest and grasslands, and the rock they used was hewn from poorly lit caves or dangerous crags. Now, we can travel easily across open landscapes, to appreciate more easily the nature of their rocky skeletons. Strolling through towns and cities, we are surrounded by rock slabs decorating buildings and walkways – often neatly cut and polished to more easily see their intricacies. Even the pebbles in our driveways contain small marvels of Earth history. It is a rocky cornucopia.

▼ **Story capsules**
Each of the pebbles on a beach or in a river is a rock sample that contains clues to the way in which it formed – a history that can stretch back millions or even billions of years.

We now also make rocks on a gargantuan scale, as concrete, brick, ceramics and other newly minted concoctions, rapidly transforming Earth's surface. Familiar to the point of invisibility to us, these invented rocks are nevertheless an extraordinary development in planetary history. And our imaginations fly yet further, aboard the cameras and sensors of the spacecraft now exploring the solar system, getting close-up views of rocks on planets and moons beyond Earth (while the first glimpses of rocks in other star systems are appearing, too).

In this book, we take a walk among these myriad rocky landscapes, near and far, to unearth the extraordinary stories – ancient and modern, huge and microscopic, hidden and exposed – that can, with a little knowledge, be read among the stones. But first, let us look at some of the most basic features of rocks.

What are rocks and what do they tell us?

Simply put, a rock is made of one or more minerals, a mineral being a specific solid chemical compound. This is a very broad definition, one that widens the scope of the subject – and this book – from some of the common understandings of the word 'rock'. Rocks are commonly thought of as hard, as in the way that people might say something is 'rock-hard'. But there is a complete gradation between, say, the loose grains of sand on a beach and an extremely tough ancient sandstone that can only be broken up with a very large and vigorously applied hammer. Old rocks are not necessarily harder, either: you can crumble some ancient sandstones between your fingers, while modern beach sands can be naturally cemented very quickly to become hard 'beach rock', which these days can enclose things like discarded drinks bottles

◄ Rock of life
The pine tree, in this age-old relationship, is supported by the rock, and gains nutrients from it, while simultaneously breaking it down into sediment, the raw material for future rocks.

▶ Inside a volcano
After the last eruption of
Thrihnukagigur, Iceland,
the magma drained away
from its shallow chamber.
Now it's possible to
descend into the heart
of this volcano.

and crisp packets as newly made 'technofossils'. And so, in reading
rocks, the best and deepest narratives are found by looking both at
ancient sandstones and at the sands on today's beaches and in today's
rivers, puzzling out how they relate to each other. They are all part of
the same grand story.

Making these kinds of links is the key to understanding our past.
We try to work out what kind of histories may be gleaned from rocks
by studying their modern equivalents. These include beach and river
sands, and also events such as volcanic eruptions and the lavas and
ashes they erupt, to help understand ancient volcanic rocks. At the
same time observing modern animals and plants and the ecosystems
they form helps us understand the fossils we find in rocks. Indeed,
most of what we consider to be the geography and biology of Earth
has relevance to the interpretation of rocks (while their chemistry and
physics may be studied as well). Understanding rocks – or trying to –
is very much a holistic exercise. That is why, in the pages that follow,
there will be a focus on processes that we can see happening on Earth
today, as well as on their fossilised remnants within rock strata. You
cannot understand one without the other.

Of course, rocks are shaped in many environments that are
inaccessible to us. The inside of a magma chamber, or the throat
of an erupting volcano, or the roots of a mountain belt 20 or more
kilometres below ground, are not places we can visit or even get near
to, even with modern technology. So, here we can turn that classic
phrase upside down, which states that preserved remains of the past
are guides to what is happening at present, far beneath our feet or in
perilously inaccessible places. Much of what we know about modern
magma chambers, or exactly what happens during volcanic eruptions
or deep within mountain belts, is based on the evidence or rocks now at

the surface, which can be more easily and safely observed, sampled and analysed. They are the witnesses to otherwise unknowable events and processes on Earth: witnesses we can see and interrogate.

Such rocks therefore act as a bridge in understanding things on Earth that are hidden from us. They are also a catalyst to further enquiry, to further explorations. For example, most sedimentary rocks on Earth were formed at the bottom of the sea. This is an environment that we land-living humans cannot easily visit, but the search for modern representatives of ancient marine strata has led scientists to explore the sea floor, by aqualung and diving suit in shallow water, and in armour-clad bathyscaphes in the deep, dark abyssal waters of the ocean floor. Scientists have gone to even greater lengths to create extreme conditions where rocks are formed: in specially built furnaces they study how rocks melt and magma crystallises, and with tiny but powerful anvils, they recreate the crushing pressures where minerals like diamond form, hundreds of kilometres below ground. We can make use of the fruits of these studies in reading the rocks of our landscapes, to help us form pictures of the kind of landscapes – and 'submarinescapes' and 'deep Earthscapes' – they represent.

Rocks can help us peer into the future of our planet, too, as yet another, contemporary reworking of that classic phrase is 'the past is the key to the future'. Particularly, as we are collectively shaping the geological future – not least by making an increasingly wide range of synthetic rocks and minerals – this aspect is something we will also explore.

The many narratives within rocks

There are many different kinds of rocks, and in this book we will have space to examine just a few of them. Rocks are classified by geologists into different categories, on a large scale into such major divisions as igneous, sedimentary and metamorphic rocks, and then into successively finer divisions. Sedimentary rocks include sandstones, mudstones and limestones, for instance, and each of these has many recognised varieties. Sandstones include pure windblown desert sandstones, clay-rich greywacke sandstones and many more.

This detailed categorisation might suggest that each of these particular rock types tells just one specific story. But this holds true (more or less) only for the specific character used in classification, such as grain size or shape in a sandstone, or how much clay it contains. In reality, the clues present in that sandstone encompass a much richer diversity of evidence to how it formed, and extend across huge reaches of time.

Thus, each grain in that sandstone was once some part of another rock – perhaps a mass of granite that was eroded away from some cliff in the distant past, or a metamorphic rock such as a gneiss that formed part

of some long-vanished mountain belt – or even a pre-existing sandstone, its grains disaggregated by wind and rain, and carried by river and sea currents, to eventually help form new sand deposits far away. To the specialist, these grains hold many clues to their long histories that can be teased out – by looking at their shape under a microscope or analysing their chemical patterns. More generally, this understanding helps us all appreciate the richness and complexity of these deeper histories, of the many different kinds of long-vanished landscapes, that can lie within one small lump of sandstone.

Then there is that geological instant when the rock took on its basic form – say, when sand grains were swept together on some beach, or in some desert, or at the bottom of the sea, to form a layer of sediment that, buried and hardened, could later become a solid rock stratum. In that brief instant are imprinted the patterns that will remain as clues to those ancient, fleeting surface environments. These clues – once we have learned to read them – allow us to reconstruct long-vanished environments.

And this is by no means the end of the story. That layer of sediment, once buried, then undergoes a series of further transformations as – buried ever more deeply – it slowly makes a journey down into Earth's crust where temperatures and pressures rise. New patterns are imprinted upon it, including those that turn it into hard rock. It can even change its character entirely to become a metamorphic rock, perhaps even melting to form new magma. Further patterns are ingrained into the rock's fabric as it slowly makes its way from the depths back towards the surface, in subterranean journeys that may last millions or even a few billion years.

When one reads a rock's history, the key question is therefore which part of that rock's long journey is being explored. Even with a single rock, there is literally no end to the pathways that can be followed. Part of this narrative richness is, of course, because rocks are immensely old, and have therefore been able to acquire a long and eventful history. Reading the age of rocks is one of the main keys to working out that history.

◄ **One rock, many histories**

The parallel striping represents layers of sediment when the rock originally formed at the surface. The irregular cracks, or rock joints, formed as the rock rose, having been buried at a great depth. Marks of erosion are visible on the surface, and are the result of weathering.

The age of rocks

Earth is immensely old – modern calculations suggest a little over 4.5 billion years old – but how do we know this? It's not a straightforward question, and took a long time to answer. The evidence had always lain concealed in the rocks of Earth, but revealing it took a long time and much effort.

The first inklings that Earth had its own planetary history that was far longer than human history began to emerge some three centuries ago, when the 'savants' of those days began to investigate the rock strata around them, as part of the general beginnings of modern science. They recognised that these rocks could contain the fossilised remains of animals and plants that did not resemble any species known to be alive at the time. This in itself was not a straightforward discovery, as much of the world was still poorly explored, and it could not immediately be ruled out that some of these fossilised organisms were not still living in a remote part of the globe. As the world's lands became better known, such survival appeared increasingly unlikely (although a few 'living fossils' were eventually found). Slowly, it was generally accepted that whole dynasties of prehistoric, long-vanished animals and plants existed long before humans appeared on Earth.

How long was that history? The approach to this question in the 19th century could not be answered directly: at that time, there was no way of dating rocks and fossils directly, in numbers of years. Some ingenious attempts were made – calculating how long it would take for the ocean to become salty, for example – but these were so flawed as to be of little use (salt in the oceans, for instance, can be removed from the sea as salt strata, as well as washed in by rivers, and so the sea's saltiness cannot be used as a measuring stick for prehistoric time). So the question was generally approached by asking: 'How many dynasties of prehistoric animals and plants were there?'. This question could be answered practically and effectively – but only after a very great deal of work by many people. Once geology was established as a science, one of the first tasks undertaken by early geologists was to systematically hammer through Earth's strata, in part for resources such as coal and iron ore, and in part to collect and catalogue the fossil plants and animals they contained.

This was an enormous task, not least because Earth's rock strata have often been crumpled and dislocated by tectonic movements, and in many places they are more or less covered by soil and vegetation that make the tracing of their different layers a most challenging task. There are places, though, where the strata appear as a neat and clearly visible layer cake – as with the Grand Canyon, for instance. From such outcrops, one of the crucial principles in interpreting rocks was worked out long ago: that older rock strata are overlain by younger ones, representing layers of sediment that have successively buried each

▶ **A history in rock layers**
The iconic strata of Arizona's Grand Canyon represent a succession of ancient environments, with the oldest at the bottom and the youngest at the top.

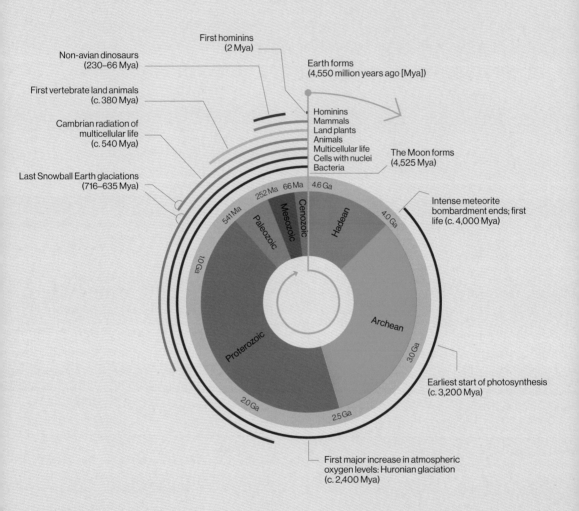

First hominins
(2 Mya)

Non-avian dinosaurs
(230–66 Mya)

First vertebrate land animals
(c. 380 Mya)

Cambrian radiation of
multicellular life
(c. 540 Mya)

Last Snowball Earth glaciations
(716–635 Mya)

Earth forms
(4,550 million years ago [Mya])

Hominins
Mammals
Land plants
Animals
Multicellular life
Cells with nuclei
Bacteria

The Moon forms
(4,525 Mya)

Intense meteorite
bombardment ends; first
life (c. 4,000 Mya)

Earliest start of photosynthesis
(c. 3,200 Mya)

First major increase in atmospheric
oxygen levels: Huronian glaciation
(c. 2,400 Mya)

66 Ma   4.6 Ga
252 Ma   Cenozoic
541 Ma   Mesozoic   Hadean   4.0 Ga
Paleozoic
10 Ga
Proterozoic   Archean   3.0 Ga
2.0 Ga   2.5 Ga

▲ **Earth's timeline**

Some major events of Earth's history are shown.
The first half a billion years have left almost no rock
record. Life, which arose at least 3,500 million years
ago, was almost exclusively microbial, and familiar
multicellular animals only became abundant a little
over half a billion years ago.

other. Called the 'principle of superposition', it remains fundamental to working out Earth history from rocks.

However, even such seemingly straightforward successions of strata as at the Grand Canyon often include huge time gaps, so that the history they contain is very fragmentary, like a book with most of the pages torn out. Those missing pages will be present somewhere else in the world, but finding them, and putting Earth's total history into proper order, is a gigantic jigsaw puzzle – one that continues, as our planet's history is enormous and infinitely intricate.

Nevertheless, by the mid- to late 19th century, the major features of Earth's biological history were clear, revealing a long succession of life forms, each appearing and later dying out, eventually leading to those alive today. This succession of fossilised life had such a generally consistent pattern that it was used to recognise and name geological time intervals. For example, the oldest rocks containing abundant fossils were used to define a Cambrian Period, most typically characterised by the fossilised carapaces of trilobites, extinct relatives of modern crabs and lobsters. Other periods were defined using the appearance and disappearance of other fossil groups, and eventually the Geological Time Scale (GTS) was assembled, which is still used today.

The main outlines of the Geological Time Scale were assembled more than a century ago, and the periods so defined – Cambrian, Carboniferous, Jurassic and others – were in place more or less as they are today. Scientists then knew that prehistoric time far outstripped human history, but did not know whether it represented just a few million years or a much longer time span. Physicists insisted on the shorter timescale, saying that otherwise Earth's interior would have cooled down completely. Geologists looking at the huge thicknesses of rock strata and the many changes to life on Earth claimed that more time was needed to accommodate all these changes to Earth's biology. In a memorable intuitive guess, the 19th-century geologist William Buckland said that the dinosaurs and marine reptiles being hauled from England's Jurassic rocks (by no means the oldest fossils) must have lived '10,000 times 10,000 years ago' (or a hundred million years).

The discovery of radioactivity in the late 19th century was the key to this conundrum. This newly recognised source of energy was what could keep the interior of Earth molten over very long timescales – and, as a very considerable bonus, it enabled geologists to date rocks as well. Once the rate of breakdown of radioactive elements was known, say from radioactive uranium into the stable element lead, then the age of uranium-bearing minerals could be worked out by analysing how much of the original uranium had decayed into lead. Using this breakthrough, it was soon discovered that Earth was billions, not millions of years old. The Geological Time Scale became ever more precisely 'calibrated' with numerical ages by analysing radioactive minerals associated with the fossils of different geological periods: minerals that formed, for instance, in lavas erupting at the same time that some of those prehistoric organisms lived. Thus, the Jurassic strata of William Buckland turned out to be some 180 million years old using this kind of analysis (and Buckland's inspired guess turned out to be not too wide of the mark).

Many types of radiometric dating have now been developed, some of which can be applied to very old strata (as with uranium and lead), and some to much younger rocks and deposits (as with radiocarbon dating, which extends back to just 60,000 years). The Geological Time Scale continues to be more precisely calibrated, allowing rocks to be dated with increasing accuracy.

The uses of rocks

It is often said of the things that we use and depend on, that if we don't grow them, then they come from the rocks of Earth. This is perfectly true – and even an understatement, for all of our crops and timber grow on soils themselves derived from the breakdown of rocks by weathering.

We see this most directly with building stones, among the earliest geological resources that humans ever used, and now a huge global business. Rocks can be used to build the frameworks of buildings and walls, and for cladding, tiles, bathrooms, kitchen floors and worktops. They help form the material framework of our lives – and also provide a cornucopia of many different types of rocks that can be examined and admired for their own sake.

Yet more visibly, natural rocks provide the raw materials for the many synthetic rocks that now surround us: concrete, bricks, asphalt, ceramics, plaster and more. These are made on such a gargantuan scale that they are a true addition to Earth's geology – and so form an array of new kinds of rocks also examined in this book.

Rocks are also the source of all the metals we use – iron and steel, aluminium, copper, titanium, vanadium and many more. These are part of our built environment, and a new feature of this planet – and via spacecraft some are now being dispersed through the solar system. A spectacular example is the beryllium mirror on the James Webb Space Telescope (JWST), which will allow us to detect distant planets and the first galaxies. We can exploit so many metals because the rocks on

▶ **Rock extraction**
There are many thousands of mines and quarries in the world (more than 12,000 in the USA alone), extracting the rock resources that we need – from sand and gravel to diamonds.

Earth are rich in metal ores – probably richer than on any other body in the solar system. So, while space entrepreneurs eye possibilities of mining the asteroid belt, the greatest number, variety and richness of metal ores may well be on the home planet.

We also build on and within rocks; the stability of our buildings depends on understanding the stability of the rocks they are built upon. Soft rocks can turn into mudflows in times of torrential downpour, and hard rocks are riven with fractures, and can generate landslides and avalanches. Some rocks can be dissolved away by water underground, leading to deep holes suddenly opening up at the surface, swallowing up houses or cars. Not all rocks are rock-solid and safe, and all these potential hazards must be understood and their locations charted, so that we may live safely among them.

Since the Industrial Revolution, rocks have grown to be the overwhelming source of our energy, too, largely through the exploitation of enormous amounts of coal, oil and gas. These fossil fuels, which have in effect trapped the energy of ancient sunshine over hundreds of millions of years, have largely powered the building of the modern interconnected world we live in. (Significant amounts of energy have come from nuclear power as well, from uranium also sourced from ores within rocks.)

Now, as the unwanted and dangerous side effects of fossil fuels – a sharp increase in atmospheric carbon dioxide levels and the global warming and ocean acidification this causes – come into play, it is to other kinds of rocks that we must look for potential answers.

▲ **Underground energy**
Rock strata deep underground may store large amounts of oil and gas that have built up over hundreds of millions of years. Such fossil fuels largely power our modern lives, but their continued use increasingly destabilises Earth's climate.

In basing our economies on renewable energy, such as from the sun and wind, very large amounts of resources will be needed, such as the rare earth elements crucial to the manufacture of wind turbines. There is the possibility of trying to dispose of excess carbon dioxide from the air, too, by pumping it into exhausted oil and gas reservoirs underground – in effect mirroring the process used to extract the hydrocarbons in the first place, using the same kind of geological and engineering skills.

Left to themselves, rocks would eventually cool Earth's climate and reverse the effects of global warming, by reacting with the carbon dioxide in the air and turning it into mineral carbonate. This is a slow process, however, which takes many thousands of years, so it cannot by itself solve our immediate climate problems.

Over billions of years, rocks have acted as our planet's main thermostat in this way, in tandem with liquid water at the surface and with the biosphere, Earth's living skin. This rock-based climate control has been key to Earth remaining a continuously habitable planet for more than three billion years. Such knowledge can help keep Earth habitable for us – and for all of life – one more reason to better understand its complex and beautiful rocky carapace.

HOW TO READ ROCKS

ROCKS: THE LITERAL FOUNDATION OF OUR LIVES

Gently warmed by the Sun, Earth has provided everything that is needed for life, not only for us as humans, but also for every other species that has inhabited the planet over its 4.6 billion-year history.

The use of rocks can be seen most easily when we walk through a city centre, admiring the variety of building stones – handsome slabs of granite, limestone and sandstone – that adorn the fronts of buildings. But the brick, concrete and mortar that make up most of our urban infrastructure are made of reconstituted rock as well – comprising a new kind of geology made in colossal amounts, a concept that will be explored later in this book. The glass in those buildings is made from particular types of sandstone, and the steel within them comes from immense iron-rich rock deposits, which were mostly formed billions of years ago, as our planet was undergoing one of its ancient transitions. The copper, lead, zinc, tin and other metals used come from different types of ores within rocks, which formed via complex pathways and processes that, in their variety and richness, are unique to our distinctive and complex rocky planet.

    The energy we use to build, heat and operate these buildings, and the transport systems that link them, also come from rocks in one way or another. The coal, oil and gas that still provide the greater part of

▼ **Capitol rock**
George Washington himself chose the original stone for the Capitol building – a sandstone formed 100 million years ago, when dinosaurs walked Earth. However, this rock soon became heavily weathered. Now Georgia marble, more resistant, forms its dazzling façade.

our energy come from rock strata; indeed, coal is a rock. For nuclear power, the uranium comes from another form of rock deposit. If we try to capture the energy from the Sun directly, we need solar panels, which also need particular kinds of mineral that we extract from rock. Our dependency here is absolute – and inescapable.

Our biological lives, and those of our fellow species, are inextricably linked with rock, too. We grow our food on soil, which is a breakdown product of rock that provides most of the nutrients for our crops. The key plant ingredient that provides carbon, carbon dioxide, is taken from the atmosphere, but before plants evolved it was exhaled from Earth's rocks, not least by volcanic eruptions. Indeed, carbon dioxide undergoes a complex exchange system with rocks, which for most of Earth's history has ensured that there has been not too much nor too little for plants as food. Critically, carbon dioxide also ensures Earth's climate stability – a stability now threatened by human misuse of rocky resources.

Therefore, the stuff that *we* are made of – calcium, carbon, phosphorus and all the other elements – has come from rocks as well, by one route or another.

Rocks are also fascinating. Now that we have learned to read their code, they tell countless stories of how Earth came to be as it is, and how it works now. In the following pages, we will explore the secrets hidden in rocks.

▲ **An extraordinary rock**
Uluru, or Ayers Rock, Australia, is made of sandstone strata that were once river sands, hardened and tilted near-vertical by earth movements a hundred million years later. These strata have unusually few natural fractures, which is why they stand so high.

EARTH: A ROCKY DYNAMIC PLANET POWERED BY HEAT

In the 19th century, a dilemma arose when early geologists began to find evidence, through many dynasties of life forms preserved as fossils and enormous thicknesses of strata, that Earth must have an almost unimaginably long history. For physicists, calculating how much heat Earth must have lost to attain its present state, where molten rock – magma – can still come to the surface through volcanoes, argued for a much shorter time span of only a few tens of millions of years.

The dilemma was only resolved later in the 19th century, when the phenomenon of radioactivity was discovered. It was quickly seen that the natural radioactivity of rocks could act as an extra heat source, to stop all of Earth's rocks from freezing completely solid, and to maintain it as an active, energetic planet for a time span now known to be over 4.5 billion years.

We can feel Earth's inner geothermal heat if we descend into a mine: the temperature goes up about 3°C (5.4°F) for every 100 metres (330 feet) farther down under the surface. Most of the heat is from radioactivity, though some is still left over from the cataclysmic formation of Earth as asteroids and planetesimals collided very early in the solar system's history, especially the powerful collision of Earth with a Mars-sized planet called Theia, which is thought to have formed the Moon from the debris flung out and then became trapped inside Earth's gravitational field. After that, Earth would have had a magma ocean 1,000 kilometres (620 miles) or so deep. And because Earth is a very large planet and rocks are good insulators, amazingly, some heat is still left billions of years later.

▼ **Heat control**

This incandescent magma, pouring to the surface near Hawai'i's coastline, is part of how Earth releases the heat that continually builds up inside it. Once cooled, it will form dark basalt lava, part of the huge mass of such rock that, at Hawai'i, rises 10 kilometres (6 miles) from the sea floor.

EARTH AS A HEAT ENGINE

This heat can be felt at Earth's surface in volcanoes and earthquakes. On a larger, slower scale, this heat drives the motion of tectonic plates at Earth's surface – hence the 'drifting' of the continents, and the uplift of mountain belts as tectonic plates collide. Indeed, one can think of plate tectonics as essentially a means for Earth to dissipate the heat that is always building up inside it. Despite the destructiveness of volcanoes and earthquakes, this way of losing heat is in fact a very regular and gentle process. On our neighbouring planet, Venus, which does not have plate tectonics, there is some evidence that the heat simply builds up inside the planet, before being released in a planet-wide outpouring of magma. Every half billion years or so, it 'resurfaces' itself in a more violent form of planetary heat release.

Given the continual increase in temperature on going ever deeper into Earth, one might think that very soon all the rocks should be molten. But there is also a continual increase in pressure that tends to keep rocks solid, and so counteracts this temperature increase. Therefore, there is a kind of competition between heat and pressure, which means that these deep-lying rocks are mostly solid, with just small amounts of melt, until Earth's molten core is reached 2,900 kilometres (1,800 miles) down.

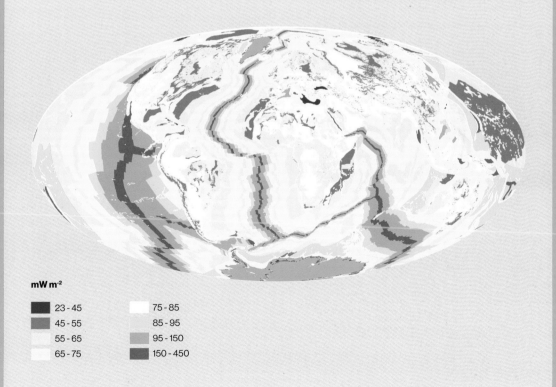

mW m⁻²

- 23 - 45
- 45 - 55
- 55 - 65
- 65 - 75
- 75 - 85
- 85 - 95
- 95 - 150
- 150 - 450

HARD INTERIOR: READING EARTH'S DEEPEST ROCKS

Mines and boreholes enable researchers to reach deeper beneath Earth's surface. The deepest mine in the world, the Mponeng gold mine in South Africa, is nearly 4 kilometres (2.5 miles) deep, while the deepest borehole, on Russia's Kola Peninsula, reached a depth of 12 kilometres (7.5 miles), about a third of the typical thickness of Earth's continental crust.

For samples of rock from even deeper down, we rely on the mechanisms of Earth itself. As the tectonic plates that make up Earth's surface move inexorably, they raise mountain belts and, in so doing, can drag huge segments of rock up to the surface from lower down in the crust or even from Earth's mantle, which lies below the crust.

Such deep-lying rocks are also brought to the surface as fragments detached by rising magmas, which can also erupt onto the surface. These xenoliths ('foreign rocks') can provide fine samples. Some of the deepest rock sources of all are those that bring diamonds to the surface. These can come from hundreds of kilometres down, in rare and powerful kimberlite eruptions.

Below that, the rocks are inaccessible to humans. Instead, they must be sensed, essentially by listening to the patterns of vibrations that have travelled through Earth. These vibrations are the shock (or seismic)

▼ **Mountain from the deep**

Slęża Mountain, seen on the skyline, is made of dense iron- and magnesium-rich igneous rocks of the 'Moho' – the transition between Earth's crust and mantle – that usually lies some 50 kilometres (30 miles) below ground. But here, tectonic forces have pushed it up to the surface.

JOURNEY TO THE CENTRE OF EARTH

It's an impossible journey for humans – but we can sense something of the character of the rock and magma layers that lie deep below us, from the way in which these layers affect seismic waves that pass through them.

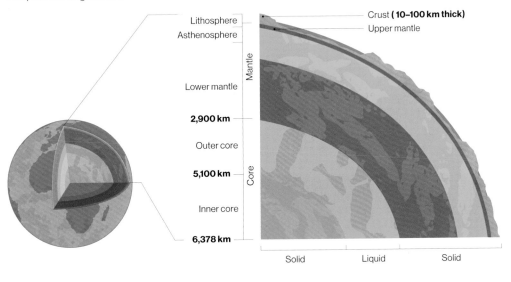

Lithosphere
Asthenosphere

Mantle

Lower mantle

2,900 km

Outer core

Core

5,100 km

Inner core

6,378 km

Crust (**10–100 km thick**)
Upper mantle

Solid   Liquid   Solid

waves generated by earthquakes (atom bombs can make them as well). Studying their path and the speed at which they travel give important clues about the nature of Earth's deepest rocks.

There are different kinds of seismic waves: some are pressure (P) waves, a bit like the sound waves that travel through air and can be heard; others are shear (S) waves, and rock particles move sideways when these shock waves pass through. Different types of rocks react to these types of waves in different ways, which helps to distinguish them.

The first major rock boundary shown by these seismic waves lies between the crust and the mantle beneath, which is made of denser material. Seismic waves show that in the asthenosphere – a section of the upper mantle – the rocks are hotter and softer. Above this is the harder and more rigid lithosphere (the uppermost part of the mantle and crust), which is broken up into Earth's moving tectonic plates. About 2,900 kilometres (1,800 miles) down is Earth's dense, molten core, which comprises iron and nickel. Only P waves can pass through it, and the study of these waves has identified a solid inner core at the very centre of Earth.

DRIVING FORCES:
PLATE TECTONICS

Unlike on the other rocky planets, Earth's exterior is not immobile. It is split into a number of tectonic plates that move slowly (about as fast as human fingernails grow) towards, away from, or past each other. This unique feature is the foundation for everything that takes place on Earth. Without it, life as we know it would likely be impossible.

The idea of such mobility emerged first in the 16th century. The way the coastlines of the Americas and Africa–Europe seemed to fit together, like pulled-apart jigsaw pieces, suggested that these continents had once been joined. This theory was forgotten, but surfaced again in the early 20th century, when more evidence was found that Earth's continents seem to have moved – at times joining, at times splitting apart. But this idea of continental drift was controversial. Most geologists refused to believe that Earth's continents could plough, as they imagined it, through the rocks of the ocean floors.

The riddle was solved later in the 20th century, when the rocks of the ocean floors (which had long been completely inaccessible and deeply mysterious) were explored by sonar, by deep submersibles and by drilling. Almost all these ocean floor rocks were found to be the volcanic rock basalt (see pages 54–61). They were also found to be geologically very young, only some tens of million years old, as compared with the ancient continents, which were a few billion years old. The youngest ocean rocks of all were seen to be those at the crests of mid-ocean ridges, where the ocean crust was continually being pulled apart and added to, as magma welled up from below.

Was the whole Earth expanding, then, like a balloon being blown up? That idea was briefly considered, but it was soon found that as fast as ocean crust was created, it was destroyed elsewhere and pushed deep back into Earth at ocean trenches along subduction zones. Zones of ocean floor creation mainly show relatively gentle volcanism. Iceland is an example: it is part of a mid-ocean ridge that has been pushed up above the ocean's surface to become dry land. However, subduction zones, and also regions where tectonic plates slide past each other (such as at the San Andreas Fault in California) are places where powerful earthquakes, tsunamis and explosive volcanic eruptions commonly take place. The 'Pacific Ring of Fire', where the Pacific ocean floor is sliding down into the mantle, is one such notorious region.

◀ **Landscape after earthquake**

The fractured landscape resulting from the deadly 1906 San Francisco earthquake along a part of the San Andreas Fault. Periodic major earthquakes are inevitable along this major fault, as it marks the boundary between the Pacific and the North American tectonic plates, which are inexorably sliding past each other.

HOW PLATE TECTONICS WORK

The key principle of plate tectonics is that the lithosphere exists as distinct, rigid tectonic plates, which rest on top of the hotter, more pliable asthenosphere within the upper mantle. The weakness of the asthenosphere allows the tectonic plates to move, relative to each other. The top globe shows the continents as they were in the time of the dinosaurs, before the Atlantic Ocean began to form.

The continents, as they parted, did not plough through the ocean floor. Instead, they are borne along on the tops of the tectonic plates, as the Atlantic Ocean slowly opens between them. The 16th-century map-maker Abraham Ortelius was right: Africa and Europe were once joined to the Americas, and modern geologists now estimate that they began to break apart about 150 million years ago.

MINERALS: THE BUILDING BLOCKS OF ROCKS

When we think of minerals, what often first comes to mind are clusters of beautiful, iridescent crystals on display in a museum. Minerals can certainly take that form, but most are not quite so blatantly eye-catching. The beauty is always there, certainly, but discerning it might require a closer look with a hand lens. And there is much more than beauty to discern, for minerals are the indispensable building blocks of rocks.

So, what is a mineral? It is essentially a naturally occurring, solid, inorganic, crystalline chemical compound. A familiar example is table salt, otherwise known as sodium chloride (NaCl). You can easily grow salt crystals, and read their characteristic cubic shape, by letting some salty water evaporate. In nature, this happens on a much larger scale where rock salt crystallises in desert environments when lakes or arms of the sea dry out. Beneath the Mediterranean, for example, there lies a deeply buried layer of rock salt up to 2 kilometres (1.25 miles) thick. This formed when the entire Mediterranean Sea became cut off from the Atlantic Ocean, some six million years ago, and dried out in the hot climate. It was a natural catastrophe, turning the sea for a period of time into a blinding-white, lifeless desert. The salt layers that formed would later be sought and exploited, where they could be reached. They were a highly prized resource, especially in pre-industrial times, before the advent of canning, when salt was the main way to preserve food.

Rock salt is one of some 5,000 natural minerals catalogued on Earth. Just a few of these combine to make the great bulk of the rocks that we see at the planet's surface. Most of these rocks are silicates, based around molecules that combine silicon and oxygen, the most common elements at the surface. Together, they make up 75% of Earth's crust. Few minerals vie with the silicates as rock formers on the surface of our planet. One exception is calcium carbonate, notably calcite, which (together with the related magnesium-rich, carbonate mineral, dolomite) makes up the limestone strata that are common in the landscape. Earth's many other minerals can be found either dotted about within the main rock types, or – especially for those spectacularly beautiful museum specimens we were thinking about – in particular concentrations such as mineral veins.

▼ Sodium chloride crystal
There is much to read in a crystal. The cubic shape is a gigantic reflection of the regular patterns of atoms within the structure. The size of the crystals is a measure of how quickly they grew (these grew quite slowly), while the glass-like transparency here suggests chemical purity.

THE ATOMIC STRUCTURE OF SILICATES

Silicates are minerals based on a particular molecular pattern – a silica tetrahedron in which one silicon atom is surrounded by four oxygen atoms (SiO_4). These link and bond with other atoms to form the crystal structures of different silicate groups.

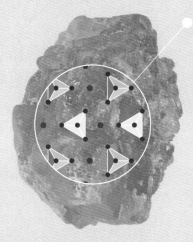

Olivine

In olivine, pyroxene and amphibole, the tetrahedra bond with atoms of iron and magnesium, and these minerals are commonly heavy and dark-coloured (though olivine is a lovely bottle-green).

Quartz

Quartz is the simplest of the silicates, with silicon and oxygen the only types of atom present.

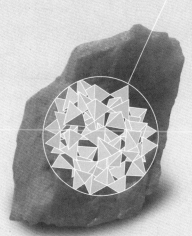

Feldspar

In feldspar – the most abundant mineral at Earth's surface – aluminium and combinations of calcium, sodium and potassium are arranged around the tetrahedra.

Mica

Silicate minerals not only exhibit differences in chemical composition, they also vary in molecular arrangements. They can be arranged as frameworks, single chains and double chains. Most strikingly, in the micas, they form sheets, a molecular pattern that is reflected in the way micas split easily into thin flakes.

FORMATION: HOW MINERALS COMBINE TO MAKE ROCKS

Only a handful of very common minerals make up most of the rocks at Earth's surface. Several dozen more minerals are also commonly encountered. But the different ways in which they can be combined, and their different proportions and patterns, give rise to myriad rock types, each of which has its own story to tell, and many of which we will explore in this book.

There are three main ways in which minerals come together to form rocks. The first is by the cooling and solidification (freezing) of magma (molten rock), to produce an igneous rock. As a certain temperature is reached, crystals begin to form within the magma, just as crystals of ice form in a glass of water placed in a freezer. Depending on the chemical composition of the melt, just one kind of crystal may form and eventually produce a monomineralic rock, as all the magma freezes (and indeed, ice may be regarded as a monomineralic rock). More commonly, several minerals form in succession, as the temperature drops, to produce a polymineralic rock. A granite, for instance, may be formed of mica, feldspar and quartz crystals, which are large because the granite magma cooled slowly underground, giving the crystals time to form. If a granite magma erupts to the surface, though, and cools quickly, only tiny crystals can form, producing rhyolite as a volcanic rock. If cooling is extremely quick, there may be no time for crystals to form, and the magma is frozen as a glass such as obsidian.

▶ **A classic granite**
The Shap Granite of northern England is made up of large crystals of pink orthoclase feldspar and smaller crystals of white plagioclase feldspar, grey quartz and black biotite mica. These crystals grew very slowly in magma that cooled and froze deep underground.

▲ **Conglomerate**
A pebble-rich sedimentary rock.

The second common way for minerals to be assembled is as sediments. Here, grains of rock or mineral can be brought together by wind and water currents at Earth's surface – one can think of the sand and pebble layers on a beach, for example – or by gravity, where mud and boulders are carried by a catastrophic mudflow. On the beach, sediment grains are continually sorted and segregated by ceaseless wave action, and so tend to be very uniform in size, while in the mudflow they may be almost completely unsorted, with boulders and mud flakes mixed up together. If these sediment layers are buried as strata, sooner or later natural chemical cements form between them to bind the grains together as a hard sedimentary rock, while some sedimentary rocks are entirely chemically formed, like layers of rock salt that form as a sea dries up. Earth has a magnificent variety of sedimentary rocks, which will be explored in Chapter 3.

Both igneous and sedimentary rocks can become deeply buried, and affected by high temperatures and great pressures, including the kind of pressures exerted when rocks are squeezed and crumpled as a mountain belt forms (where tectonic plates collide). As these temperatures and pressures rise, new minerals form, and the rocks – while staying solid – begin to change and become different kinds of metamorphic rocks. A mudstone (once a layer of mud), for example, when increasingly heated and squeezed, will form a slate that will transform (as the crystals get bigger and new minerals form), into a schist and then a gneiss. The minerals here are arranged in patterns that give clues to the forces that formed them.

All these types of rocks are bound together in a history that may be represented by the rock cycle.

▲ **A typical sandstone texture**

The desert weathering picks out the inclined strata showing this hard rock was once loose sand being sculpted into dunes by currents of wind or water.

▼ **Rock transformations**

Below is a metamorphic rock, a gneiss, its mineral bands formed by recrystallisation at high temperatures and pressures.

THE ENDLESS ROCK CYCLE: FORMATION, DECAY AND RENEWAL OF ROCKS

Look up at the Moon with binoculars, and you can see rock patterns that have been virtually unchanged for billions of years. This is because the Moon is geologically almost dead. Earth, by contrast, remains highly active, driven by its internal heat. Its constant motion is expressed through the unique mechanism of plate tectonics: continents divide and then reassemble, and ocean basins are formed and then destroyed.

Earth's rocks, caught up in this ceaseless activity, are forever being formed, broken down and transformed. Indeed, almost nothing is left from the first billion years of its history, a time equivalent to those ancient, still-visible foundations of the Moon. This geological activity can be described as a rock cycle that operates over timescales of many millions of years.

The cycle might start with the formation of igneous rocks from magma, the most primordial kind of rock. Once at Earth's surface, igneous rocks are eroded, and broken into fragments, such as occurs when storm waves strike rocky cliffs along a coastline. Rocks are also chemically attacked, or weathered, by rainwater, which is a dilute acid because of the carbon dioxide dissolved within it. Under this attack, many igneous minerals decompose, their molecular lattices breaking down to form new sedimentary minerals such as clay minerals, the chief ingredient of mud. Minerals that are more resistant to chemical attack, such as quartz, are released as grains, while some of the material of the igneous minerals is carried off in solution, ultimately to add to the salt in the sea.

Of these broken-down residues of igneous rocks – known as sediment – some stay on the land surface to form soil. Much of the sediment, though, is quickly carried away by rivers, which flow into lakes or the sea. There, sediment layers are deposited, piling up to form thick successions of strata, especially where parts of the crust are continually subsiding. Once this process has started, the enormous weight of the sediment causes the crust to sink even further, causing yet more strata to accumulate, often to a thickness of several kilometres.

The heat and pressure at these depths first helps lithify the soft sediment layers into hard sedimentary rock strata. And then – especially when caught up in mountain-building episodes – these sedimentary rocks (together with igneous rocks) begin to be altered by more heat and pressure, forming metamorphic rocks.

At great depths and high temperatures, these metamorphic rocks begin to melt, and so form magma. From that magma, igneous rocks will form, and so the cycle begins again.

THE ROCK CYCLE

Transformations are continually taking place within Earth's rocky crust, driven both by weathering and erosion at Earth's surface, and by the effects of heat and pressure deep underground.

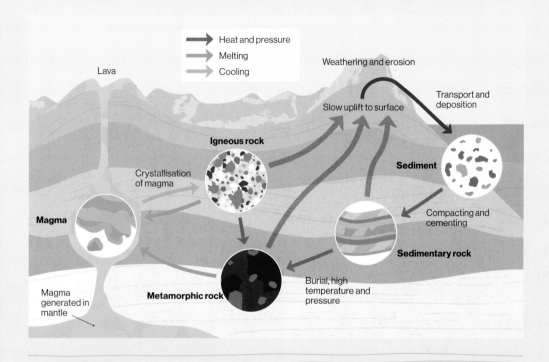

Heat and pressure
Melting
Cooling

Lava

Weathering and erosion

Slow uplift to surface

Transport and deposition

Igneous rock

Crystallisation of magma

Sediment

Magma

Compacting and cementing

Sedimentary rock

Magma generated in mantle

Metamorphic rock

Burial, high temperature and pressure

◀ **Rock erosion – close-up**
Sedimentary strata being wave-eroded on Aberystwyth beach, Wales, to form pebble deposits in the beginning of this local rock cycle, while finer sediment is washed farther out to sea.

WHERE TO SEE ROCKS:
FROM URBAN TO NATURAL SETTINGS

One of the many delights of taking an interest in rocks is that they are visible almost everywhere – far more so now than in the time of the Stone Age. In those days, only some of our ancestors lived in rather dark, rocky caves. Those that lived on the plains were, more often than not, separated from the rocks that lay beneath their feet by thick layers of vegetation and soil. However, today rocks have been excavated in enormous masses for our buildings and other constructions, and the once endless forests have been widely felled and cleared to make agricultural fields. In these open soils, it is easy to find and examine pebbles and rock fragments.

It makes sense to start our rock journey of discovery close to home. While few of us have magnificent rocky crags nearby, most of us have access to soils in our local parks and gardens, and these typically contain pebbles and natural cobbles within them – natural rock samples, each of which, properly examined, can show patterns that give clues to their long and often dramatic geological histories. Some of these neatly sized rock samples have come from nearby, from the rocks beneath the soil. Others have been dragged tens or hundreds of kilometres by glaciers during the ice ages, and can illuminate the geology of a whole region – often from just one collection made in one

◀ **Urban rocks**

Towns and cities are never far away from the rocks they are built on, and these might be seen in natural crags, or in human-made excavations and road cuts. Carefully approached, the rocks thus laid bare can reveal the deepest history of our urban areas.

◀ **A classic mountainous landscape**

Mountain belts, as shown here in the Italian Alps, are classic places to admire large expanses of rock. But be careful – studying rocks close up in such regions can be difficult and perilous.

place. Yet others have been brought in by humans, to make gravel paths and drives. However they arrived, rocks can include fascinating textures, structures, minerals and fossils that open a window into past worlds, right on your doorstep.

Pay a visit to the nearest town or city centre, and the range of human-brought rocks expands massively, often in a most spectacular fashion. This is where a wide range of rocks can be admired and pondered, from those that make up the ornamental fronts of shops, banks and restaurants, to the materials used to create sculptures or line public walkways. Even better, these have often been cut and polished, so that their textures and structures are clearly visible, whether on a large scale, as when standing back from some majestic building, or when peering more closely at microscopic details. There are also vast amounts of human-made rocks, of course, which have their own considerable interest (see Chapter 6). As a bonus, somewhere in the city there may be a museum containing beautiful rock specimens and fascinating information about them. A city, in short, is something of a geological paradise.

Of course, one can go into the mountains or along a rocky coast and see rocks in their natural surroundings. But they are often not so easily examinable as on the decorative rock slabs of a city (and their position may well make them dangerous, too!). However, the beauty in seeing them in their natural order enables the examination, not only of the details of the rocks, but how the different rock masses relate to each other. You can then more easily work out what kind of extended history they reveal together. Each landscape has its own history, and discovering this is like solving a crime mystery, but on a far grander scale.

▼ **Desert rocks**

In desert areas, such as here in Arizona's Vermilion Cliffs, there is little soil and vegetation to obscure the rock textures, as in this case of ancient sandstones.

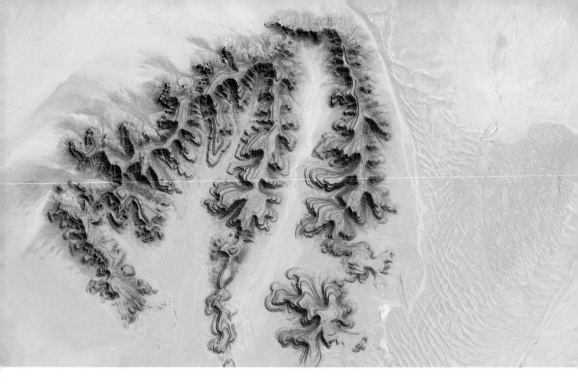

A QUESTION OF SCALE:
FROM PLANETS TO GRAINS OF SAND

Rocks can be examined at all scales, from a macro level of a whole planet or moon, to a microscopic and submicroscopic level with grains of sand. From any of these perspectives, stories of the rocks' long histories can be drawn out.

These days, images of the planets and moons of the solar system are widely available, thanks to space exploration. Of all of these, that of Earth is in some ways the most difficult to interpret, as this is a dynamic, living planet with weather systems, vegetation, soils, oceans and cities that cover much of its rocky foundations. Nevertheless, it is now easy to find satellite views of mountain ranges and rocky deserts, for example, that can be endlessly explored on a computer screen.

Coming down to Earth, one can puzzle over the landscapes seen, to try to understand the rocks beneath – a particular geological skill that will be explored further. And within these landscapes are crags and cliffs where rocks are exposed, and can be examined more closely. For this, the most important analytical tools by far are your eyes. For the most part, you will not need that legendary piece of geological equipment, a hammer! When hammering, it is all too easy to do more

▲ **Rocks in satellite view**
A satellite image of the Namib Desert in Namibia shows a fern-like pattern of ancient rock formations, in which the large-scale, flat-lying strata can clearly be seen, picked out by erosion. Around these rocks the pale yellow areas are modern active dunefields of windblown sands.

harm than good, and to ruin perfectly good rock faces. In particular, many rock faces in the landscape have been exposed to the wind and rain for decades or centuries – and in that time weathering has often picked out and accentuated subtle but textural features of the rock that are clues to its origin, but that may be hard to see on a fresh rock face. These include such things as the outlines of ancient dunes, fossils, or blocks within ancient lava flows. To preserve those rock faces for future generations, it is best to look, not touch!

To look even more closely, there is an extraordinary, indispensable and ultra-low-cost piece of equipment that can reveal whole new hidden worlds: the hand lens. With a magnification of only around x10, this can show an extraordinary amount of detail that is simply out of reach of the keenest naked eye. It can show the nature and arrangement of crystals in igneous rocks, the shape and composition of grains in sedimentary rocks, the fine anatomy of fossils, the textures of mineral deposits and a whole host of other rock clues. Small enough to carry in your pocket, it can, as a bonus, also show you things like flowers and insects in beautiful close-up as you walk through the rocky countryside. There is a small trick to using a hand lens properly: the lens should be held close up against the eye and the rock specimen brought up to it (or the head should be bent towards the rock face). Tying a ribbon to the lens will make it harder to lose while exploring.

To look even more closely, you can go higher-tech (and more expensive), and use a binocular microscope for greater magnification – and, even more revealingly, cut 'thin sections' of rock (so thin they are translucent) to examine under a polarising microscope. This is specialist, professional work – but luckily there are libraries of such images freely available on the internet.

▲ **Close-up rock**
View of a sandstone as seen through a microscope – the shape and size of the sand grains can be clearly seen, as well as of a small pebble in the middle of the field of view.

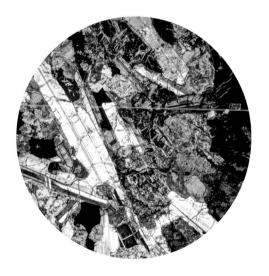

▶ **Microscope view of rock**
To examine with an optical microscope, a rock is cut into a 'thin section' – a translucent slice just a thousandth of an inch thick. Shining polarised light through it gives the minerals their different telltale colours.

TOPOGRAPHICAL CLUES:
DECIPHERING THE LANDSCAPE

In many parts of the world, especially where the climate
is mild, and soils and vegetation develop easily, rocks can
be difficult to see directly. They are there, not far under
your feet – but they remain tantalisingly hidden from
view. This is especially true with what geologists call 'soft
rocks' – the youngest strata of all, formed a geological eye
blink ago during the last ice ages (see pages 150–151), and
even afterwards, after the climate warmed, in the last few
thousand years. These recent strata have only rarely been
lithified into hard rocks, and so do not stand out as crags.
Yet, the stories they contain can be just as fascinating and
important as those of the 'hard rocks' of more ancient times.

Geologists often must search hard for where rocks might be exposed.
But there is another skill to bring into play, and that is in analysing
the whole landscape to work out the arrangement of the hidden rocks
beneath. Here, one tries to develop X-ray eyes, in order to peer beneath
green hillsides and plains, and reconstruct the rocky skeleton that lies
beneath. Geologists call this 'feature mapping', because they use the
evidence of topographical features – ridges, valleys, changes in the angle
of a slope – to puzzle out the geology beneath. It is a skill anyone can
develop, and has the added advantage of involving nice walks through
the countryside (though this can also be done by looking at satellite
images of landscapes on a computer screen).

Some of the most straightforward examples can be found among
the soft rocks. For example, the flat floodplain of a river marks where
the river channel has, over thousands of years, meandered from side to
side across the valley it occupies, constantly changing its position and
leaving the sediments of the river – pebbles, sand and mud – in its wake.
The edge of these particular soft rocks, the recent river sediments, is
therefore where the flat floodplain adjoins the valley slope that rises to
higher ground. These geological boundaries can be easily spotted in
every area where there are rivers.

Similarly, in mountainous terrain, one can look for the difference
between rough, high, craggy ground where the ancient hard rocks are
near the surface, and the smooth and gentle slopes that adjoin them,
marking where the much softer, younger sedimentary deposits (such
as those left by glaciers during the last ice age) lie immediately beneath
the land surface, covering the hard rocks beneath.

HOW ROCKS FORM TOPOGRAPHY

Within the tilted strata being eroded here, the bed of hard rock is more resistant to erosion, and so forms a ridge with a steep scarp face, behind which is a gentler dip slope. Geologists use these kinds of clues to trace geological strata across landscapes – even where the rocks themselves are not exposed.

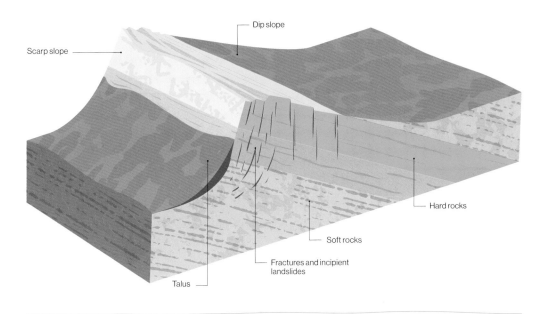

Scarp slope

Dip slope

Hard rocks

Soft rocks

Fractures and incipient landslides

Talus

Within the hard rocks there is a very common topographic feature with alternations of hard, resistant strata (such as sandstones) and softer-weathering strata (such as mudstones). The hard strata form ridges (called 'scarp features') that can be tracked precisely across the countryside – even when the rocks themselves are not exposed at the surface – to map out these sandstone layers, while the low ground between these ridges marks out where the intervening mudstone layers are.

A variety of other such telltale features are present in the landscape, some of them spectacular, such as the neck of an ancient volcanic vent that forms a steep hill in Edinburgh, Scotland. It takes time and practice to recognise them all – but it is a most enjoyable and satisfying skill to develop.

ROCKS FROM MAGMA

DEEP HEAT:
THE MAKING OF MAGMA

Igneous rock is the end point of magma's remarkable journey, which begins deep underground, where rock is heated by the natural radioactivity of rocks. Because Earth is so large, and rocks are such good insulators, the heat builds up until the rocks begin to melt.

But there is more to melting than just heat. If we could take a journey several hundred kilometres underground, we would be deep in Earth's mantle, at a temperature of some 1,700°C (3,092°F). On Earth's surface, rocks at this temperature would be quite molten, but the crushing pressure in Earth's mantle ensures they remain a dense crystalline solid. This deep, solid, incandescent rock is nevertheless flowing on long, exceedingly slow journeys through the mantle.

When such rock rises to about 200 kilometres (125 miles) below Earth's surface, it is still very hot, but the pressure is lower. In this zone, melting begins. Only a small proportion of rocks in the mantle liquify, forming little magma droplets. These droplets have a different chemistry than the mantle rock that remains solid, typically with less iron and magnesium and more silicon and aluminium. This kind of 'stretching out' of the chemistry of igneous rocks in the melting process is one of the ways by which the great range of igneous rocks, with their different chemical and mineral compositions, is formed on Earth.

These magma droplets, with their altered chemical composition, are often less dense than the remaining rock, and so rise through it. The routes are initially slow and tortuous: magma can take many thousands of years to work its way upwards towards Earth's surface.

When magma begins to ascend through the more brittle, cooler rocks of the crust, it often follows cracks and fractures. Sometimes it will gather in magma chambers around 5 to 10 kilometres (3 to 6 miles) down, where the magma accumulates in masses of hundreds or even thousands of cubic kilometres. From these, the magma, impelled by some tectonic change, or by new inflow to the chamber, can reach the surface as volcanic eruptions.

At these depths, the magma comprises not just molten rock, but also dissolved gases, such as steam and carbon dioxide. The high pressure usually keeps these gases within the magma, just as carbon dioxide remains in a fizzy drink until the bottle or can is opened. Once the magma nears the surface, these gases are released in the eruption that follows.

THE ANATOMY OF A VOLCANO

Volcanoes are the end result of long and complex journeys made by magma deep below ground. The nature of those journeys, and the way in which they can change the composition of the magma, determines the kind of eruption that will take place, and what kind of volcanic rocks form. Wherever the magma stalls and solidifies underground, different kinds of plutonic rocks are formed.

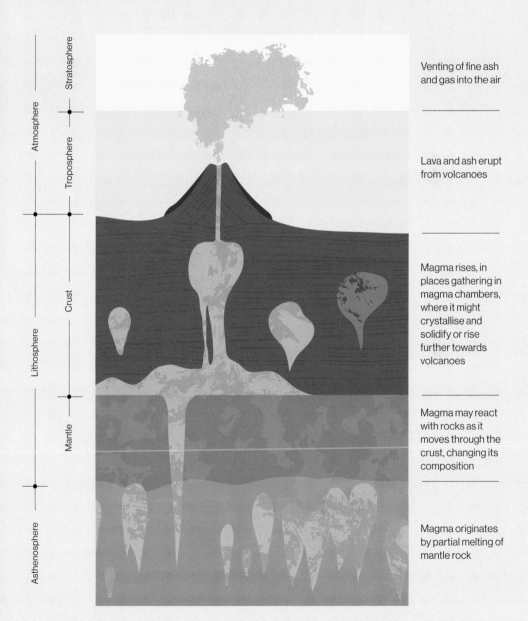

Stratosphere

Troposphere

Atmosphere

Lithosphere

Crust

Mantle

Asthenosphere

Venting of fine ash and gas into the air

Lava and ash erupt from volcanoes

Magma rises, in places gathering in magma chambers, where it might crystallise and solidify or rise further towards volcanoes

Magma may react with rocks as it moves through the crust, changing its composition

Magma originates by partial melting of mantle rock

COOLING DOWN:
THE FORMATION OF PLUTONIC ROCK

Not all rising magma makes it to the surface. If the journey to the surface is too slow, it may cool and freeze along the way. If magma encounters surface rock masses that are less dense than it is itself, it becomes easier for that magma to simply stay underground and lift the rocks above a little higher, than to travel further upwards. Much magma, therefore, cools and solidifies underground, to form igneous rocks called 'plutonic' (named after Pluto, the ancient god of the underworld), or 'intrusive' (because the magma intrudes into other kinds of rocks).

The surrounding rocks, especially if they are some kilometres below ground level, are warm, and they also insulate the magma. Entire magma chambers can cool and solidify underground in this way, exceedingly slowly – over tens or even hundreds of thousands of years. As temperatures gradually drop, chemical ions in the magma begin to stop moving freely, then join together to form the molecular frameworks of crystals of different forms of minerals. With this very slow cooling, there is plenty of time for crystals to grow large, as seen in granites and gabbros.

Very large and spectacular crystals can grow where an additional factor comes into play. Magmas that are water-rich become loose and runny, and so the ions can move through them more easily towards growing crystals, which can therefore grow larger. Such spectacularly large crystals can be seen in rocks known as pegmatites, which form

▼ **Mineral treasure troves**
The magmas that give rise to pegmatites concentrate rare elements, and this can produce spectacular crystals, like this titanite from the Alps.

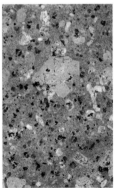

◀ **Slow and fast crystallisation**
The large crystals here of quartz, pink feldspar and dark amphibole grew slowly in a magma chamber, with the rest of the rock cooling quickly to a finely crystalline groundmass after eruption towards the surface.

next to bodies of granite, from the last, water-rich pockets of cooling granite magma.

On the other hand, magmas that are injected into narrow cracks and fissures of cold rocks high in the crust (or those that erupt on the surface) form thin layers that cool quickly. Here, there is little time for the ions in the rapidly cooling liquid to move very far. Under these conditions, very many small crystals (which can be hard to see, even with the use of a lens) form to make a fine-grained igneous rock. In some cases, if the magma is very viscous and cools very quickly, there may not be time for crystals to form at all, and the rock simply becomes a frozen liquid – a natural glass, such as obsidian. It is hard to see through such a natural glass because of impurities, and so one would not use it for windows. However, because it breaks to give very sharp edges, it was often used by our ancestors to make tools and weapons.

Some of the most attractive igneous rocks are those where slow cooling began deep down, where large crystals began to grow before the magma suddenly erupted to higher, colder levels. These early-formed crystals then become set as large phenocrysts within a mass of much smaller, later, fast-grown crystals. These rocks, called 'porphyries', are often used as a decorative stone, their beauty reflecting the drama and complexity of their origin.

▲ **Granite mountain**
Goat Fell on the Isle of Arran, Scotland, is a granite pluton that formed when the adjacent part of the Atlantic Ocean was opening, 55 million years ago. It now rises so high because it is made of harder rock than the more erosion-prone strata into which it was intruded.

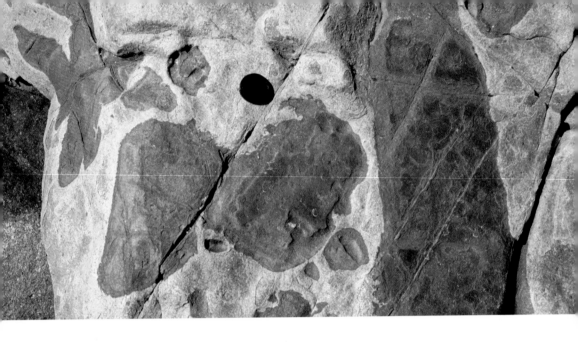

CLOSE-UP: EXAMINING GRANITES AND GABBROS

Granites and gabbros (and the rocks that are intermediate between them, called 'diorites') make a useful and beautiful introduction to igneous rocks. Having cooled slowly from large masses of magma deep underground, the crystals have grown large, so that they can be seen even without the use of a hand lens (although with such a lens, you will see yet more detail). Just walking around a town or city centre, you are especially likely to see many examples of granite, beautifully polished for easier examination.

Most granites, which are silica-rich and usually pale or brightly coloured, are made of just a few minerals. Making up most of the rock are crystals of feldspar, which are either colourless or white (if they belong to the sodium- and calcium-rich plagioclase feldspars) or an eye-catching red–pink (if they are a potassium-bearing orthoclase feldspar). There is usually a fair amount of quartz as well, which is also typically a clear to colourless mineral, so might be confused with plagioclase feldspar.

How do you tell them apart? This is one of the first tasks of the budding geologist, for these two are among the most common minerals on Earth. Looking at a granite is a perfect way to start distinguishing them. The feldspar here is usually in box-shaped crystals, which on

▲ **When magmas meet**
A rock made of two kinds of magma – pale granitic and dark gabbroic (now as finer-grained dolerite rock) at Tuross Head, Australia – frozen and solidified while in the act of mingling together underground.

broken edges often show a kind of 'staircase' fracture pattern. This is because the mineral tends to break along flat mineral cleavage planes that reflect its internal molecular pattern. The quartz, by contrast, does not have such mineral cleavage planes: it fractures along curved surfaces, like glass. Another difference is that the feldspar is prone to slow chemical decay by weathering, then looks more whitish and less translucent, beside which the quartz (which resists such weathering) by comparison appears grey. Here begins a distinction process you will make for as long as you look at rocks.

Granites also usually contain a scattering of darker minerals – usually either mica (with crystals shaped like hexagons, which split perfectly along internal cleavage planes into innumerable flat sheets) or a dark amphibole or pyroxene. Hidden in this granite will also be tiny amounts of minerals such as zircon (terrifically useful for geologists as it can be used for measuring the age of the rock, by measuring how much the tiny amounts of uranium within it have radioactively decayed) and apatite, a phosphate of calcium. To see these, a microscope is usually necessary.

Gabbros, by comparison, are darker, denser rocks. They also contain a lot of feldspar (usually the whitish plagioclase form), but little or no quartz. Instead, much of the rock is made up of those dark minerals that only occur as sprinklings in granites, if at all – i.e., the amphiboles and pyroxenes. Amphiboles and pyroxenes can be quite hard to tell apart: the amphiboles form six-sided crystals, with cleavage planes at 120 degrees, while pyroxene crystals are more eight-sided, with cleavage planes at right angles – but you need to look hard to see this.

Some granites and gabbros include spectacular varieties. Granites can contain veins of pegmatite, with impressive crystals of a large range of minerals. Both granites and gabbros can include orbicular kinds, with round, layered crystals like magmatic hailstones. There is a rich world of variety to explore when examining just these two rocks.

▲ **Earth's most common mineral**

The silicate mineral feldspar is the most abundant mineral in Earth's crust, and is a major ingredient of both granites and gabbros.

▼ **Granite types**

(Left) In this granite, one can see large crystals of feldspar (pink and white), quartz (grey) and dark mica (black). (Right) This granite shows spectacular orbicular texture.

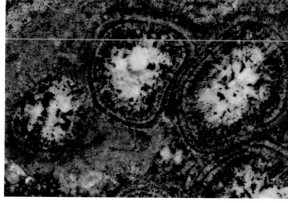

FRAGMENTS OF DEEP EARTH: XENOLITHS

As magma rises towards Earth's surface, it can detach fragments of rock (as xenoliths) and mineral (as xenocrysts) from great depth and carry them with it. These are witnesses to the journey it has taken, and some can represent parts of deep Earth that humans cannot, and probably never will, reach and experience.

Some of these rock passengers, though, are strictly local. In a granite magma chamber, fragments of the rock wall of the chamber can be broken off and become incorporated into the granite mixture, which is usually a viscous mush of melt and slowly growing crystals. Such xenoliths can be quite common around the edges of a large mass (a pluton) of granite. When the rock wall is made of a contrasting kind of rock – say a dark-toned gabbro or basalt – then these xenoliths can be easily visible, for example, on the kind of polished granite slabs one can see on building façades.

Some xenoliths can stay more or less intact, especially when caught up in some 'cool' granite magmas, which can be not much more than 500°C (930°F). But if the magma is much hotter, such locally caught-up rock fragments can melt, and thus become part of the magma. Large amounts of hot basaltic magma can even melt substantial amounts of rocks of the crust that they come into contact with. Those melted crustal rocks are themselves turned into magma; being more silica-rich than basalt, they can go on to become granites.

▼ **Gabbro xenolith in granite**

This dark xenolith from California's Sierra Nevada is surrounded by the pale granite that, as magma, tore it away from its original position in the bedrock.

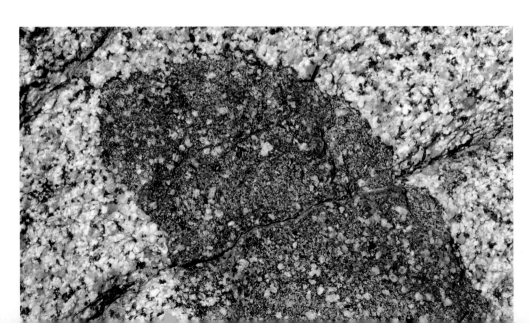

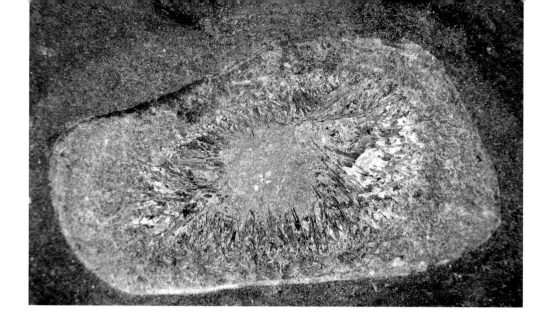

Igneous rocks originated at great depths, often below the crust and within Earth's mantle. But they only indirectly represent those deep-lying rocks, being the most easily melted parts of them, and therefore different in their chemical and mineral composition. However, fragments of those original mantle rocks can be carried up to the surface with the magma, especially within fluid, and particularly with dense and relatively fast-moving magmas, such as those of basalt. Within these dark rocks can be found lumps of a beautiful greenish rock: a peridotite, made up almost completely of the mineral olivine (which in its gem form is known as peridot). This is thought to make up much of the upper parts of the mantle.

▲ Altered xenolith
This striking xenolith, found inside ancient Precambrian rocks in Ontario, Canada, shows a radial pattern of crystals of the mineral actinolite. It represents a heavily altered fragment of the lower crust or mantle, caught up in ascending magma.

▶ Mantle xenolith
The beautiful bottle-green colour of this xenolith of peridotite shows that it is mostly made of the mineral olivine: it is a typical piece of mantle rock, erupted with its surrounding lava onto Earth's surface in Arizona.

MAGMA INJECTIONS:
SILLS AND DYKES

Some of the most eye-catching examples of igneous rocks in the landscape are igneous dykes, where magma has ascended along near-vertical fractures in the crust, and then has become frozen in place within them. These fractures may once have been the conduits for the magma to reach the surface high above, to spread out as lava fields; or they may have been dead ends in which the magma simply came to rest deep underground.

▼ **A wall of igneous rock**
The dark basaltic rock here takes the form of an igneous dyke, and represents basaltic magma that flowed along near-vertical fractures. It contrasts markedly with the pale rock that it intruded into.

Whichever was the case, they were generally the result of the crust being stretched by tectonic forces, so that it breaks apart, like pastry being pulled apart. In rocks, myriads of closely spaced, near-parallel fractures can open up, to act as magma pathways. Once the overlying rock has eroded, after many millions of years, then these fracture fills of long-solidified magma, termed 'igneous dykes', often form multiple dyke swarms. These can be seen on the present-day land surface, on rocky shorelines and other places where they are not hidden by soil and vegetation.

Igneous dykes often form spectacular features, especially where the igneous rock is harder than the rocks into which the magma was intruded (which might have been, say, relatively weak sedimentary rocks). The resulting walls of hard rock – long, but often just a metre or two across – can have the appearance of being human-made. On a beach, you can be fooled into thinking that they are artificially constructed breakwaters.

Viewed more closely, the rock can show telltale signs of its history. Near the boundaries of the fracture, the magma has often cooled a little more quickly when it came up against the relatively cold surrounding rock, so its crystals may not have grown as large as those in the middle of the dyke. And the escaping heat might have visibly 'cooked' the adjacent rock, to form what geologists call a 'baked margin' to the dyke. How deep was the magma when it was intruded, at the point where you are looking at it? If you see little ancient gas bubbles frozen within the igneous rock, it must have been near the surface, so the pressure on the magma decreased enough to allow the gas to begin to bubble out. If you don't see gas bubbles, it will likely have been deeper. There are many such detective games to play with these rocks.

In some places, the magma may have stopped ascending, and then begun to spread out sideways within the surrounding rock mass. This typically happens when overlying rocks are less dense than the magma, so it is easier for them simply to be pushed upwards than for the magma to rise higher. This kind of magma behaviour is also encouraged where there are near-horizontal planes of weakness in the rock (such as the planes between thick beds of sedimentary rock). The resultant masses of igneous rock form igneous sills, which can also form spectacular landscape features such as cliffs and plateaus (especially when the magma has contracted as it cooled and solidified), to form a fracture pattern called 'columnar jointing' that can look like gigantic, natural organ pipes.

▲ **An igneous sill**

A spectacular sill in Edinburgh, Scotland, formed when basaltic magma intruded underground between layers of near-horizontal strata. The basalt lava seen today has well-developed cooling joints.

MAGMA REACHES THE SURFACE:
VOLCANIC ERUPTIONS

When magma breaches Earth's surface, it can do so peacefully and quietly, allowing onlookers to stand a modest distance away and safely gaze at the spectacle. Or, it can break through the surface so swiftly and violently that people may not even be safe a thousand kilometres away. Which of these outcomes arises depends, largely, on what kind of magma is present.

If the magma is runny, most of the gas dissolved within it will escape as the magma rises. This is the case with most basalt magmas, which are relatively low in silica and rich in iron and magnesium. When such magma breaks the surface, it is mostly degassed, and so will pour out as lava – white-hot at first, and soon chilling, hardening and darkening. This is the kind of eruption you often see on nature documentaries on TV.

▼ **A small volcanic eruption**
Runny basaltic lava fountaining during an eruption at Kīlauea volcano's Puʻu ʻŌʻō vent, Hawaiʻi.

Such lava flows can damage property, but they are generally not too dangerous to human life – although the volcanic gases can be toxic.

Silica-rich magma, however, is viscous (because the silica molecules polymerise within it), meaning it flows less easily and traps gases within it. These gas bubbles turn the magma into a sticky and rapidly expanding magma foam that, on breaking the surface, fragments explosively, producing a violent volcanic eruption – essentially a powerful explosion that can be sustained for hours, until the magma supply is exhausted. The fragmented magma foam quickly hardens into pumice, much of which is subsequently pulverised into ash, and carried high into the sky by the upward-billowing gases. These are catastrophic volcanic eruptions, and a single one can devastate a whole region. Caught up in an eruption like this, you would be in pitch darkness, with ash and rock fragments falling out of the sky.

If you were particularly unlucky, you might find yourself in the path of a turbulent mixture of hot gas, pumice and rock too dense to rise into the sky, which instead pours out of the volcano mouth like milk boiling over a pan, speeding along and hugging the low ground. These are the fearsome pyroclastic flows, or nuées ardentes ('burning clouds'), lethal to everything they come into contact with.

A less apocalyptical outcome will be produced by very viscous lava, which, if degassed enough, can inch up a volcanic vent as a near-congealed mass, extruding to form a solid lava spire perhaps a few hundred metres high. Such giant rock columns soon collapse under gravity, though, sending an avalanche of debris down the flanks of the volcano.

TYPES OF VOLCANISM

These graphics give an outline of the different kinds of volcanism that take place on Earth. Not to scale.

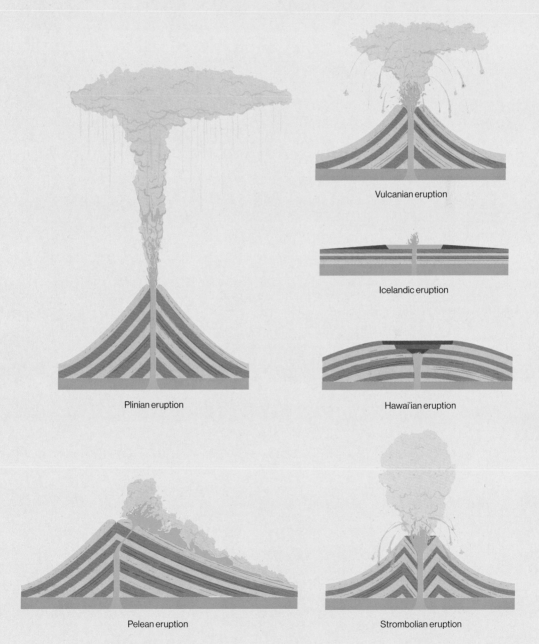

Plinian eruption

Vulcanian eruption

Icelandic eruption

Hawai'ian eruption

Pelean eruption

Strombolian eruption

EXPLOSIVELY DIVERSE:
THE TYPES OF VOLCANO

Volcanoes have enormous variety. Some are so small and neatly formed that you could almost fit them into a back garden. Some are large enough to be the highest mountains in the world (measuring from the ocean floor, where their base is); these often have smaller volcanoes dotted across them. Others are simply jumbles of broken rock. The deadliest ones may sometimes be barely noticeable as you travel past, looking like mere depressions in the ground.

The simplest and smallest (and most common) volcanoes are aptly called 'cinder cones' (or 'Strombolian cones'). They can be just a few tens of metres high and a few hundred metres in diameter; some of them form from just a single eruption, lasting a few months to years. The 'cinders' (or scoria) are lumps of bubbly basaltic magma with just enough gas to power repeated small eruptions, showering these particles a little way into the air, for them to fall nearby and build up a steep-sided cone.

▼ **Miniature volcano**
A small Strombolian (or 'cinder') cone on Hawai'i. Its sides are made up of layers of air-chilled bubbly lava fragments (scoria) that have cascaded down its side during eruptions. Scoria fragments often have complex, twisted shapes from their trajectory in the air.

The largest volcano in the world is Mauna Loa, on part of the Hawai'i island chain. It is a shield volcano – very broad, with gentle slopes (some are so gentle, they are barely noticeable at all), but one that built up from the sea floor, 5 kilometres (3 miles) below sea level, to eventually rise to 4 kilometres (about 2.5 miles) above sea level. It is mainly made up of many basaltic lava flows (and is dotted with cinder cones as well).

The 'classical' volcanoes, such as Mount Fuji in Japan, are usually made of more silica-rich magma, which erupts both ash and lava to build up the large, steep cones that are most associated with volcanoes. However, these are enormous, steep masses of fractured rock and loose ash, in environments prone to earthquakes and further eruptions. It is their fate, sooner or later, to collapse in enormous avalanches to form rubble piles that spread out across the landscape. On many island volcanoes in particular, giant 'bites' seem to have been taken out of their outline where sectors of them have collapsed into the sea, often causing tsunamis (the resulting avalanche deposits on the sea floor may cover a greater area than the volcano itself).

In the most powerful, catastrophic eruptions, most or all of such a volcano can be destroyed, blasted into the air or collapsing. In such super eruptions (like those that have taken place in the Yellowstone volcano, in prehistoric times), entire magma chambers, comprising a few thousand cubic kilometres of magma, can be erupted explosively and spread as ash layers across extensive landscapes. As the magma is expelled, the ground above simply collapses to form a depression tens of kilometres across, often lined with steep cliffs facing inwards. In the centuries that follow, these calderas often fill with water, to form a lake that looks peaceful – until the next super-eruption. Luckily, these events are rare – there has not been a 'Yellowstone'-scale eruption in recorded human history. But one will happen, sooner or later.

▲ **A classical volcano**
The Mount Fuji volcano in Japan has the classical cone shape, with a crater at the top. The oval scar on its lower slopes at the bottom of the image gives a hint of its ultimate fate: it is a small landslide scar, a reminder of the instability of these steep mounds of lava and ash.

PILLOWS AND FOLDED ROPE:
LAVA FLOWS

When magma freezes, it sometimes retains a flowing 'liquid' appearance. This occurs when very liquid lava (usually basalt lava) chills and freezes. This solidified surface is often folded and contorted by the liquid lava flowing beneath it, to resemble carelessly folded rope. Indeed, it is often called 'ropey lava', and the pattern in which the 'ropes' are bent and folded in an ancient lava body can reveal the direction in which the lava flowed.

The front of such a lava can advance slowly and gently enough that it is possible – with care – to get close and observe the process. The frozen lava front may briefly appear almost motionless – but every now and again, a red-hot bleb of magma will break through the cooled surface as a tube-like protrusion, which then quickly cools and darkens before the next one bursts out alongside it. The whole lava creeps forward in this way by innumerable tiny breakouts, to produce what is known as 'pahoehoe lava' (a Hawai'ian term, as such lavas are common on the islands).

Basalt lava erupting underwater often forms similar structures (cooling against water rather than against air), to produce what is known as 'pillow lava': these lava protrusions look like a giant stack of rock pillows lying on top of each other (in reality, they are typically more tube-like than pillow-like). Much of the ocean floor forms like this, a process that can be observed (with the help of a bathyscaphe) at the mid-ocean ridges where new crust is being formed – although, farther away from the mid-ocean ridges, on older sections of ocean floor, these pillow basalts become buried under sediment.

Many lavas, however, look much more like thick piles of angular, jagged rubble. These form where the magma is more viscous, and breaks up the chilled carapace into large blocks, which are then slowly carried or pushed along by the still-flowing magma beneath. Such a lava flow looks like a pile of rubble and boulders with steep sides and front: the blocks are slowly pushed forward as though by a giant bulldozer, tumbling over each other noisily as the lava keeps the whole mass growing and advancing. Different varieties of such lava are called 'a'a lava' and 'blocky' lava. The fresh surface of such a lava can be a perilous place to cross on foot – it is very easy to break an ankle or trap a leg between the blocks. Best to admire from a small distance!

▶ **Pahohoe lava**
The cooled surface of a pahoehoe lava with a classic 'ropey' texture. The preserved pattern of the 'ropes' suggests that the still-liquid lava beneath was flowing from right to left.

▶ **A'a lava in the making**
Carried by the incandescent, still-liquid lava are the myriad jagged blocks characteristic of a'a lava flows – formed by the continual solidifying and breaking up of the lava top.

TRAPS IN TIME:
GIANT LAVA LANDSCAPES

Some of the most extraordinary landscapes in the world are formed by enormous flows of basalt that come to rest one above the other, eventually piling up in large masses that can be several kilometres thick and cover many thousands of square kilometres. The resulting landscape, shaped by millions of years of subsequent erosion, often takes on a stepped appearance, where each giant 'step' is the eroded edge of a single lava that may be ten or sometimes hundreds of metres thick.

Such landscapes (once called 'trap' landscapes) can be seen in many parts of the world where such enormous pulses of basalt magmatism broke through to the surface. There are fine examples in the Deccan area of India, and around the Columbia River in the USA. The terrain is made yet more spectacular where the lavas show beautifully regular columnar cooling joints, as in the Giant's Causeway of Antrim, Northern Ireland.

▼ **Building up basalts**
This kind of succession of basalt lavas, some of which have characteristic columnar jointing, when multiplied many times over, is what builds up the 'trap topography' of Earth's biggest lava outpourings.

Some of these impressive landscapes hide darker secrets. The Siberian Traps are a prime example; they represent an extraordinary outpouring of magma that took place 250 million years ago. At that time, an uprising 'plume' of mantle material impinged on that part of Earth's crust, generating a huge pulse of magma that, in about two million years, caused some 4 million cubic kilometres (960 million cubic miles) of basalt to pour out, in flow after flow. It now seems certain that the volcanic gases that were released with the lavas caused a major global warming event, acidified and de-oxygenated the oceans, and dispersed toxins. It is believed that this caused some 95% of animal and plant species to die out, in the greatest mass extinction event to ever have affected Earth. Other basalt outpourings have been linked with major environmental change, too, at different times in Earth's history.

As a final twist, these enormously thick and extensive lava flows used to be referred to as 'flood basalts', with geologists envisioning that they poured out rapidly, in fiery, fast-racing torrents across the landscape, submerging it almost instantaneously. Later, geologists looked more closely at modern lava flows, and saw that many of them crept forwards only slowly, for instance as 'pahoehoe' lavas. Behind the slowly advancing lava front, the solidified crust on top of the lava could be seen to be inching upwards, as if it was being jacked up. Here, it is inward-flowing lava that does the jacking, expanding the whole body of the flow as though it was a balloon (only filled with lava, and not with air). Indeed, this slow, gentle, process is now known as the 'inflation' of a lava, and is likely responsible for most of the 'flood basalts' of the world. However, this does not take away from the dangers posed by huge outpourings of basalt.

▲ **Lava outpouring**
Along with the lava flowing out of this active volcano on Fagradalsfjall, Iceland, toxic gases are emitted, with compounds of sulphur, fluorine and chlorine, as well as carbon dioxide. With large eruption episodes, these can have severe environmental impacts.

VOLCANIC ASH LAYERS:
ASH FALL

In the more explosive and catastrophic volcanic eruptions, typically associated with stiffer, more silica-rich magmas, much of the erupted material froths up into pumice, which – together with rock fragments from the disintegrating volcano – is then carried as volcanic ash high into the air, often to heights of 20 kilometres (12 miles) or more, so it can enter the stratosphere. It does not reach that height by the explosive force of the volcano alone. Such 'ballistic' propulsion would shoot the ash fragments only a kilometre or so upwards.

The additional energy comes from the giant heat engine that is the volcano, creating a superheated cloud of material that – even when weighted down with the pumice and rock – can be less dense than cold air, and so rises upwards as the dark, billowing, turbulent updraught of an eruption column. When it reaches a level in the atmosphere where the surrounding air is so thin that it cannot rise any more, the eruption column spreads out sideways in a giant mushroom cloud. From this cloud, the pumice and rock fragments then fall those 20 kilometres (12 miles) or so back down to the ground, to carpet it in a layer of white pumice, where dark rock fragments are dotted like raisins in a fruitcake.

Such pale ash layers can be seen around most of the world's dangerous explosive volcanoes. They have a distinct appearance. The pumice fragments, after their long journey up into the sky, then along

▼ **A volcanic ash fall deposit**

These beautifully size-sorted angular pumice fragments in Tenerife, Spain, which fell perhaps 20 kilometres (12 miles) out of the umbrella cloud at the top of the eruption column, are typical of ash fall deposits from a major explosive eruption. There are also a few darker and more dense volcanic rock fragments.

PLINIAN ERUPTION

These explosive eruptions are called 'Plinian' after Pliny the Younger, who wrote about the eruption of Mount Vesuvius, Italy, in 79 AD. The eruption column is driven extremely high by the immense heat released from the volcano.

When the eruption has finished, and the magma chamber is emptied, much of the superstructure of the volcano can collapse into the cavity that is left, forming a volcanic caldera.

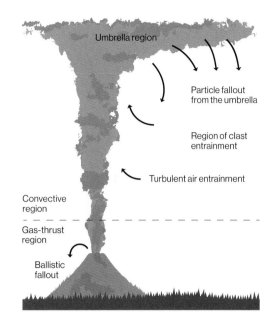

Umbrella region

Particle fallout from the umbrella

Region of clast entrainment

Turbulent air entrainment

Convective region

Gas-thrust region

Ballistic fallout

the spreading eruption cloud and down to the ground, are typically marvellously well sorted, as if they had been through some kind of huge industrial sieve, while the adjacent rock fragments, being denser, are smaller. Near to a volcano, the deposits of a truly big eruption drape the landscape, and can be up to a few metres thick: they may bury trees, houses and cars. Such a single-deposit eruption can show layering, with some finer and more coarse layers that track the history of the eruption as it waxed and waned across several hours or days. The same layer can be followed across the ground away from the volcano: it becomes progressively thinner, while the rock and pumice fragments within it become progressively smaller – and yet it will preserve the same pattern of layering that reflects the history of that eruption, just on a finer scale. It is a bit like a fingerprint and allows the 'ash fall' layer to be distinguished from those of other eruptions from the same volcano, each of which will have its own distinctive pattern of internal layering. This is the kind of evidence that allows volcanologists to reconstruct, by such forensic study, the history of a volcano, and thus to help discern its hazards for the people who live around it.

VOLCANIC ASH LAYERS:
ASH FLOW LAYERS

In 1902, after the tragic eruption of Mount Pelée that struck the once-bustling town of St. Pierre on the Caribbean island of Martinique, the realisation came that the ash from a huge explosive volcanic eruption did not always rise high in the air, simply to fall back to the ground. Here, the eruption column was too dense to loft upwards, but instead spilled out of the volcano as a ground-hugging cloud of turbulent hot gas, pumice and rock fragments. This deadly effusion raced at express-train speeds downhill to overwhelm St. Pierre, destroying many of its buildings and killing all but two of its some 30,000 inhabitants.

The nuée ardente-type eruption that overcame the island is nowadays more usually known as a pyroclastic flow. A common part of many explosive eruptions, the 'ash flow' layers, or ignimbrites, that it leaves behind, have been recognised by geologists worldwide around volcanoes both active and extinct. Also termed 'ash hurricanes', they are a terribly dangerous form of eruption. However, volcanologists now understand how they form and travel, and this helps to prepare communities in volcano-prone areas for the kind of hazards they pose.

◀ **Destruction wrought by a pyroclastic flow**

▲ Damage from Mt. Pelée in St. Pierre, Martinique, in the Caribbean, 1902 (left) and from Mt. Merapi in Java, Indonesia, in 2010 (above). In both cases little ash remained after the pyroclastic flow had passed through.

Ash flow layers are strikingly different from the beautifully size-graded ash fall layers. In ignimbrites, an unsorted mixture of fine ash, pumice and rocks is rapidly dumped out from the bottom of the searing pyroclastic flow as it speeds along the ground. Some of the blocks can be impressively large, having been carried or dragged along the ground by the dense, speeding current. The pumice fragments here are unlike the angular pumices found in ash fall deposits, but are usually swiftly and effectively rounded by bouncing off other particles in the hot, swirling current. Also unlike ash fall layers, which evenly drape low and high ground alike, ignimbrites pile up in valleys to thicknesses of hundreds of metres – they usually do not cover higher ground, as the dense pyroclastic flows follow the lowest parts of the topography around the volcano.

Ignimbrites can be so hot when they are deposited that the ash particles fuse together, as the whole deposit compacts down under its own weight. This forms a rock called a welded ignimbrite, in which the originally strongly curved pumice bubble-wall fragments of fine ash particles are flattened and pressed together, while the large fragments of pumice are flattened into streaks (geologists label them 'fiamme', from the Italian word for 'flames', due to their shape). In some ignimbrites, this process goes so far that the rock effectively re-melts, and looks very similar to a lava. In the aftermath of eruptions like these, the landscape seems to have been enamelled with volcanic glass.

▲ **Ignimbrite rock**
An ignimbrite, shown here in close-up, is typically a poorly sorted mass of fine ash, with rock and pumice fragments, contrasting sharply with the beautifully size-sorted deposits of volcanic ash falls (see pages 62–63).

THE DIAMOND VOLCANOES:
KIMBERLITES

Diamonds are, for many, precious symbols of eternity. That seems appropriate, given that they are the hardest and most incompressible natural mineral on Earth – and for good measure, the best conductor of heat and of sound waves. The journey they have endured from their origin, deep within Earth, to be explosively erupted up to the surface, is one of geological high drama, too.

Diamonds form at a depth of hundreds of kilometres inside Earth – some as deep as 1,000 kilometres (620 miles) down, in the lower mantle. Their raw material is a more ordinary form of carbon, such as the organic matter within black mudstones, carried down by subducting tectonic plates. At those depths, it is compressed and heated over enormous timescales – many diamonds are billions of years old – to form the durable cage-like molecular structure of diamond crystals.

The diamonds then have to make their way to the surface. The main propellant for this likely has the same carbon-based source that made the diamonds themselves – the gas carbon dioxide – together with water in superheated form. If enough of this propellant accumulates, it can begin to drill its way upwards through the solid mantle rock, advancing perhaps as fast as a human can run, and carrying with it lumps and fragments of mantle rock – including diamond crystals. By the time it nears the surface, this explosive mixture accelerates to jet plane speeds, punching a pipe-shaped hole in the crust that may be a kilometre or more across, and launching high in the sky in a kimberlite eruption.

Kimberlite eruptions would have been truly awesome, but none have occurred in recorded human history – most took place tens or hundreds of millions of years ago. What is left are the kimberlite pipes drilled through the crust, which are filled with rock and mineral fragments brought up from great depth, including diamonds. Most of this igneous mixture has long since been altered by weathering to form

▼ **Chunks of a mighty diamond**
The nine major uncut stones split off the rough Cullinan diamond.

a 'blue' clay, within which are the near-indestructible diamonds that today are so avidly sought. Quarrying out the blue clay can produce impressive, deep, circular mines, such as that at Kimberley itself, in South Africa.

Diamonds contain stories that unlock the mysteries of Earth's interior, at depths that no human could ever reach. When diamonds crystallise, they can trap within them tiny fragments of the other minerals that grew around them, and the rigid cage of the diamond crystal can preserve them on their long journey to the surface. One diamond found in Brazil about 500 kilometres (310 miles) below the surface included a speck of a mineral called ringwoodite that was found to contain water in its structure. From this, it was estimated that Earth's interior contains at least an ocean's worth of water. Another diamond found in Botswana in 2021, sourced from even deeper levels, contained a new high-pressure, never before seen mineral christened 'davemaoite': once liberated from its diamond cage by laser, it survived only a second before expanding and transforming into glass. Diamonds may be forever, but some minerals within them only survive for as long as they remain within a diamond's protective embrace.

▼ **Ghost of a kimberlite pipe**
The hole in the ground after mining of the Mir kimberlite pipe in Yakutia, Russia, between 1957 and 2004. The mine is 525 metres (1,700 feet) deep, and is 1,200 metres (nearly 4,000 feet) in diameter.

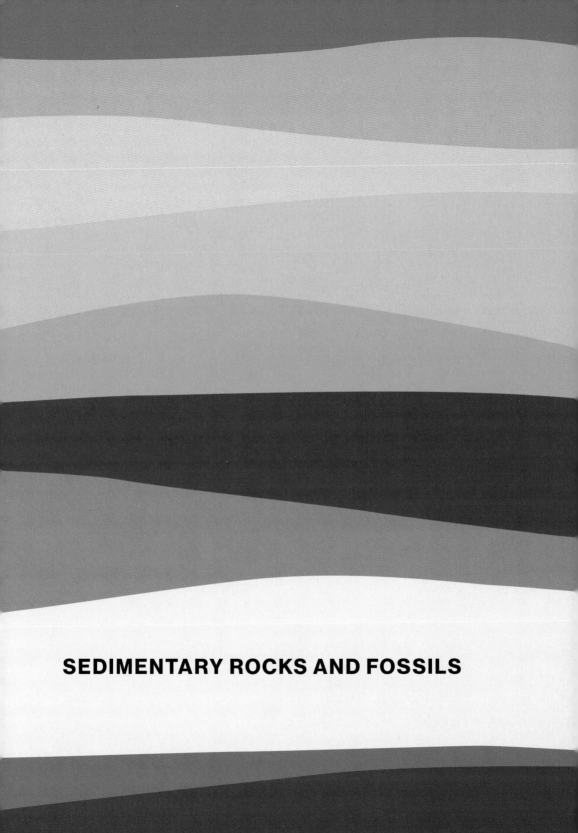

SEDIMENTARY ROCKS AND FOSSILS

GRADUAL OBLITERATION:
EROSION, WEATHERING AND DECAY

It is easy for us to take the constant slow destruction of our landscape by wind and weather for granted: a symbol, one might think, for the impermanence of all things. But gaze at some other planetary bodies, and one sees landscapes that are almost eternal. On the Moon, for example, the bright areas seen on a clear night are highlands that still bear the signals of being pulverised by meteorites four billion years ago, the Moon's lunar 'seas' are lava flows that welled out onto the surface two to three billion years ago. Earth – a planet marked by constant decay – is the unusual body, but this very decay has led to the constant renewal of Earth's store of rocks.

Rocks on Earth are constantly being destroyed, physically or chemically. Their physical destruction can take many forms. Sometimes it is simply the effect of gravity, exploiting weakness in the rock masses. Thus, volcanoes – masses of lava and ash piled high above the surrounding ground and so inherently unstable – can simply collapse (see page 57). The same process can happen more widely, where the compressional forces of plate tectonics lift mountain ranges several kilometres above sea level. The most spectacular – and deadly – forms of destruction are landslides. An avalanche from a mountainside in Longarone, Italy on 9 October, 1963 sent 0.25 cubic kilometres

▼ **Landslide scars**

This mountain stream has cut into the adjacent steep slope, forming small landslides, and the streamflow is washing the sediment debris towards the sea.

▲ Coastal erosion

The wind energy gathered by waves across vast expanses of ocean is focused on erosion of the adjacent coastlines, seen in magnificent, and constantly receding, cliffs. The beach sands below them represent part of this moving train of debris, washed and sorted by the constant wave action.

◄ Rock weathering

Chemical weathering has effaced the features on this 14th-century statue – said to be of King Aethelbald – in Crowland, eastern England.

(0.06 cubic miles) of rock debris plunging into the reservoir below; the floodwater sweeping over the dam killed 2,000 people. Ancient examples can be far larger: in the collapse of a mountainside in Bonneville, Oregon, 14 cubic kilometres (3.4 cubic miles) of rock debris blocked the Columbia River, to form the 'Bridge of the Gods'.

The Columbia River has already washed much of this natural dam away, while rivers are carving valleys and gorges across landscapes worldwide. Storm waves gather wind energy from thousands of square kilometres of ocean surface and unleash it on the coastline, to wear away cliff lines. In deserts, winds fling sand grains against rock surfaces, and slowly abrade them, while in icy landscapes, glaciers grind away at the rocks beneath.

Partnering this large-scale physical destruction, chemical weathering takes place to break up rocks. Water is a supreme solvent, and its effect is heightened by carbonic acid (from carbon dioxide in the air) and humic acids (from decaying plants). Many minerals – not least those of igneous rocks formed at high temperatures – succumb to the cold, wet, corrosive conditions at Earth's surface, and their molecular frameworks are dismantled. Signs of such rock decay are seen everywhere – even walking through a village or city centre, where rock slabs on a building crumble, or passing through a churchyard , where words carved into old gravestones are in the process of gradually being effaced.

Under this chemical attack, high-temperature igneous minerals such as olivine and feldspars decay into minerals that are more stable at Earth's surface, such as the clay minerals that go into muds. Some of their chemistry is dissolved into rain or river water, to ultimately flow into the sea as ions and make it salty. Some minerals, such as quartz, resist chemical attack and are released as grains; they are then carried away as river sediment. All of these ingredients make up the next generation of rocks.

▲ **Lava cave**
This 'lava tube' on Hawai'i is the space left when liquid lava
flowed out from beneath a quickly-hardened lava carapace.

Caves

For many of us, caves evoke the lives of our
distant ancestors and the legends of the
underworld – and in truth, they provide a
unique and direct access to a world below
ground, which would otherwise remain closed
off to us. Contrary to Jules Verne's classic
novel, however, we cannot reach the centre
of Earth, nor anywhere near to it: at a great
enough depth the extreme pressure would
crush a person flat. But before those depths
are reached, caves can tell many stories.

Some caves are original features of the rock.
For instance, ancient lava flows sometimes
contain lava tubes, which form when a very
runny lava develops a solid crust: once the lava
has flowed out, a tunnel some metres wide
and quite possibly tens of kilometres long
remains. The sides of such a tunnel might show
horizontal 'bath marks', created by a decreasing
level of flowing magma. Additionally, 'lavicicles'
can hang from the lava tube ceiling. Once
cooled, such lava tubes can be inhabited by
humans and other animals.

Most caves, however, are produced by flowing
water, which slowly dissolves rocks and
produces underground spaces. The main
rock type involved here is limestone, which
is dissolved by naturally acidic rainwater.
The percolating water at first widens cracks
and fissures in the rock and, as dissolution
continues, eventually produces complex cave
systems. This type of landscape is known
as karst, characterised by surface limestone
pavements marked by the criss-cross pattern
of these widened joints. The cave systems
within them can be huge and, where the roofs
of these eventually collapse, vertical-sided
gorges in the landscape are produced.

It is not just chemical dissolution
that takes place in limestone caves. Some
of the calcium carbonate taken from the
limestone can be re-precipitated within the
caves as speleothems, which can take the
form of stalactites and stalagmites (see page
opposite). These not only provide spectacular
underground scenery but in the course of

their growth, over many thousands of years, they become rocky witnesses to changes on the surface, even climate change. This is because the mineral structures have absorbed chemical clues to the changing environment above, faithfully transmitted down by the percolating waters that nourished them.

Caves offered shelter to all kinds life long before humans appeared on the planet. Fossil cave systems have been found that date back many millions of years, and which contain rich stores of fossilised remains of animals. Ice-age caves have been found to contain the skeletons of bears and hyaenas, as well as the bones of humans' ancestors. Furthermore,

cave paintings left by our ancestors offer a fascinating glimpse of how they viewed the world around them.

Of course, it is not only limestone that can form karst landscapes. In its rock form, gypsum dissolves away even more easily, and underground gypsum layers can also contain cave systems. As gypsum rock is soft, many of these have collapsed and now form chaotic 'broken strata'.

THE COMPLEXITY OF CAVES

Intricate patterns of solution and deposition control the labyrinthine structure of caves.

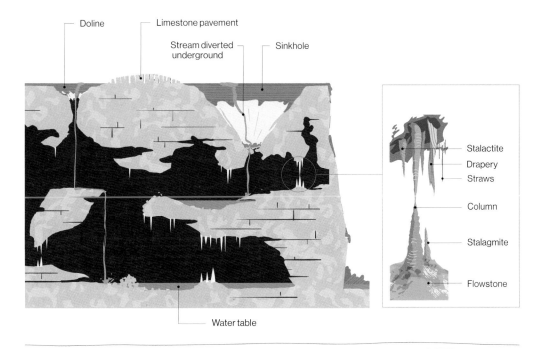

Doline · Limestone pavement · Stream diverted underground · Sinkhole · Stalactite · Drapery · Straws · Column · Stalagmite · Flowstone · Water table

FROM RIVERS TO THE SEA: THE ENDLESS SEDIMENTARY CONVEYOR BELT

On Earth, the landscape is constantly changing, constantly refashioning itself. On a human timescale, the hills around us may look eternal, but they are constantly eroding. Much of the sediment released from them first forms the soil that covers the rocks. That soil then moves slowly downslope under the action of gravity, its movement helped along by the rain washing over the soil surface and the animals burrowing within it.

Eventually, the sediment finds itself in rivers, which are channels not only for flowing water, but for the pebbles, sand and mud that form a kind of moving carpet at the bottom and sides of the river channel. Here is the start of a long process of separating out the different kinds and sizes of sediment. Much of the mud travels steadily along, more or less constantly, the fine clay flakes suspended in the flowing water.

INEXORABLE GRAVITY

Although the surrounding landscape may seem stable from day to day, wherever there is a slope, the soil and rocks beneath are constantly, very slowly moving, under the pull of gravity. This diagram shows some of the resulting clues in the landscape.

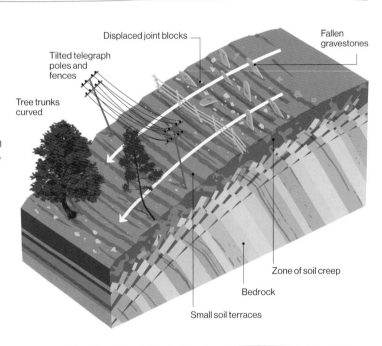

Displaced joint blocks

Fallen gravestones

Tilted telegraph poles and fences

Tree trunks curved

Zone of soil creep

Bedrock

Small soil terraces

Pebbles and cobbles stay mainly in the bottom of the channel, where the current is strongest. Even then, they are mostly not moving, as they are too heavy for normal river flow to shift. Only when the river is in spate does it have enough force to dislodge pebbles and cobbles, at which point they clatter downriver until the flood flow subsides, settling into new positions a little farther along the river's course. The sand grains, part rolling and part bouncing along the riverbed, separate out from the mud and pebbles. They make intermittent halts up on the inside bends of the river, where they pile up to form curved sand bars, which shift position as the river slowly changes its course.

As well as this natural sorting into different grades of sediment, each travelling at different speeds along different paths, the sediment particles themselves are changed by the travel. This is most obvious with cobbles and pebbles. Repeatedly colliding with each other in their flurries of movement, their sharp edges are worn away and they become smoothed. When swept in a current, they often pile up, overlapping like a pack of cards pushed over sideways. Or, one large cobble can get stuck in the riverbed, with smaller pebbles piling up behind it. The sand grains also become more rounded, only more slowly, as being smaller their collisions are less energetic. The clay flakes can change at a microscopic level as well, sometimes clumping together by electrostatic attraction, sometimes dispersing, sometimes absorbing or releasing different chemicals from the water.

When the river enters the sea, the travel can be halted as the sediment builds up to form large deltas. Or, the travel can continue along the coast, along beaches and spits, to places that are particularly good at sorting and smoothing pebbles and sand grains from the constant wave action. The sediment may also carry on travelling, washed by storm waves and tides into shallow seas. Alternatively, it may continue on, propelled by gravity into the deep water of the oceans. In all these places, layers of sediment can build up – and this is the start of new rock strata.

▲ **A meandering river**
The position of the river channel is constantly shifting – sometimes splitting, and sometimes leaving abandoned segments as U-shaped oxbow lakes. The sediment it carries remains as myriad curved sandbars in the wake of the migrating channel.

Where, today, are the rock strata of the far future forming? Mostly, out of sight. Much takes place on the sea floor, a place that is difficult for us to reach. But there are places in today's landscape where we can begin to see how these processes work.

In a river, one can peer through the flowing water to see pebbles in the channel, or sand swept up onto the bank. But these are only tiny samples of these modern river strata. For the whole picture one needs to look across at the entire low, flat floodplain, which may be kilometres wide for a big river. Beneath the flat floodplain surface lie deposits that can be several metres thick, laid down by the river over the past several thousand years, as it has constantly shifted its position.

These modern strata might be quite temporary, at least in geological terms. Over the coming millennia, the river might erode all these sediments away again and carry them into the sea. Or these strata might be preserved, buried, changed into hard rock and be a witness of today's river, for many million years to come – just as today, geologists can find beautiful examples of ancient river deposits, some of them billions of years old. Which of these alternative futures will happen depends on what the underlying crust is doing.

If the crust is slowly rising tectonically (like most mountainous areas are doing today), then everything at its surface, including such recently-formed river deposits, will inexorably be eroded away and carried towards the sea. If the crust is slowly subsiding, those river deposits might be buried under more river deposits, which are themselves buried, to form thick masses of strata.

These kinds of tectonically sinking areas are Earth's main strata factories, and because Earth is tectonically active, it has amassed enormous amounts of strata through its long history. It is, in a very real sense, the strata planet.

Some of the biggest modern strata factories are where big rivers are pouring huge masses of heavy sediment into a place where the crust is already sinking. There, the extra weight of this sediment make the crust, which is pliable at this scale, sink even further. In America, the Mississippi River flows into the Gulf of Mexico at New Orleans, and the sediment deposited there over millions of years has created the Mississippi Delta, which continues to sink through its own weight. Beneath the streets of New Orleans the sediment layers accumulated in just the last 10,000 years are 100 metres (more than 300 feet) thick.

The same is true of the landscape around Venice, Amsterdam, Shanghai, Lagos and many other places. These modern cities are just concrete veneers constructed on gigantic stratal layer cakes several kilometres thick. One day, the remains of those cities will form their own distinct stratum within that layer cake.

▼ **Delta sedimentation**
Part of the Mississippi Delta, which grows as sediment from the US interior builds up in and around the delta channels. Pale plumes of suspended fine sediment can be seen, and this sediment will drift farther, to settle in deeper water.

RIVER JOURNEY

A river carries water and sediment from the high ground where it originates, down to the sea. Along the way, it changes its nature, and each part of it has its own particular pattern of erosion and sedimentation.

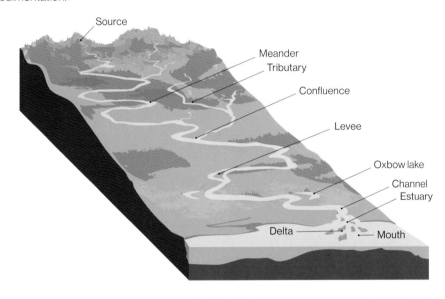

Source

Meander

Tributary

Confluence

Levee

Oxbow lake

Channel

Estuary

Delta

Mouth

◀ **Channel of a long-vanished river**
These 300-million-year-old strata in Nova Scotia, Canada, reveal a former river channel preserved in sandstone.

TRANSFORMATION:
HOW SEDIMENT BECOMES HARD ROCK

The layers of loose sediment seen on beaches, riverbanks and the soil in gardens can all be transformed, over time, into solid rock strata that can be so hard the geologist's hammer simply bounces off them. How this happens is not always straightforward – especially as regards the time element. There are sand layers from the days of the dinosaurs that can be dug into by a spade, and mud layers as old or older that have stayed soft enough to make bricks from. There can be beds of tough 'beach rock' on some modern coastlines as well, which contain plastic bottles and empty crisp packets, and so have clearly hardened within just a few decades. How can one explain this kind of contradiction?

▲ **Iconic conglomerate**
England's 'Hertfordshire Puddingstone' is the remains of a 55-million-year-old pebbly beach, and is now an extremely tough rock – the sand and pebbles are bound together by a natural silica cement.

In each case, it is necessary to think through the particular rock's history with regard to the processes that turn soft sediment into hard rock.

One factor is chemistry: in many sedimentary strata, the individual particles, whether they are pebbles, sand grains, mud flakes or fossil fragments, are held together by some type of natural cement. One common cement is calcite (calcium carbonate). Its source is often fossil shells buried together with the sediment. Once buried, these could be dissolved in the groundwater that filled the spaces between the sediment grains. Then, if the chemical conditions changed, the calcium carbonate could crystallise from the water around the grains, to bind them tightly together into rock. This kind of cement forms very quickly, if the ingredients are there, as in the modern 'beach rock'.

Most examples of such cement are hard to see: pretty well invisible to the naked eye and hard to see even with a hand lens, they will only be revealed properly with a microscope. Sometimes, though, if the sedimentary particles are large – say, pebble-sized – and the cement forms thick coats, then its nature becomes clear even to the naked eye.

There are other kinds of cement. One is silica, which can originate from the remains of fossils with silica-based skeletons such as 'glass sponges' or microscopic diatoms (a type of single-celled alga), or can come from the grains themselves. Iron oxides can also act as a cement – these typically give the rock a striking red colour. Rocks like these must form in oxygen-rich conditions – either on land or in the floor of a shallow, well-oxygenated sea. Another kind of cement is made by clay

minerals, though these do not make for a very strong cement, and a rock bound together like this can often be crumbled by hand.

As well as chemistry, pressure can help bind sedimentary particles together under the kind of conditions where rocks are buried very deeply, with a few kilometres' thickness of other rocks lying on top of them. At these depths, quartz grains lying on top of each other begin to fuse together in a process called pressure solution. Pressure is highest at the points of contacts of the grains, and here the silica begins to dissolve, the ions going into the groundwater that bathes the grains. Once in that lower-pressure situation, the silica tends to precipitate out again and forms a tough coating around the grains, binding them together.

▲ **Natural cementation on a grand scale**

In this 'megabreccia', in Titus Canyon, Death Valley, USA, the large angular blocks of dark limestone (that may have been broken up underground by tectonic movements) have been firmly cemented into their current position by white calcite.

PEBBLES, BOULDERS AND MONOLITHS:
THE LARGEST SEDIMENTARY GRAINS

The more technical term for a 'grain' is a 'clast' – a separate
rock or mineral fragment that makes up part of a 'clastic
sedimentary rock'. So how big can a clast get? It can range in
size from a microscopic clay flake, to a kilometre-sized block
called a 'mega block'.

Mega blocks can form when a mountain, or a large part of the sea floor,
collapses in a rock avalanche. Huge rock slabs can then slide downwards
like gigantic sledges, within thick masses of more finely broken-up
debris. An individual mega block from such an event, exposed in ancient
strata on a hillside, can appear like an enormous crag.

On a smaller scale, boulders are a common type of clast within
some kinds of sedimentary strata. These are commonly found in rock
avalanche strata, but boulders can also be carried by powerful floods
of water in steep terrain, such as those pouring off a steep hillside
in a deluge. Ancient boulder-rich strata of this kind are common
in mountainous areas, and represent part of the history of their
destruction by the forces of erosion. Boulders of volcanic rock can

▼ **A boulder left by ice**
Glacial erratics –
boulders carried by
glaciers and ice-sheets
of the ice ages, and then
left scattered across the
countryside after the ice
melted – are common
in many areas that have
been glaciated.

also be carried by pyroclastic flows, and are a typical feature of the ignimbrite strata that result (see pages 64–65).

Cobbles (brick-sized rock lumps), and especially pebbles, are even more common in sedimentary strata, to the extent that rocks can be made up of them. These are mostly rock fragments, though some can be made up of just one mineral, notably quartz. In working out the history of a pebble, its shape provides important clues. Angular pebbles form when fragments that break off some rock face are buried before there is a chance for their rough edges to be smoothed. This can happen in deposits like scree, where splinters of rock tumble down to cover a mountainside. Such a deposit, once fossilised, is called a breccia. So-called fault breccias can form underground, where rocks are broken up as a tectonic fault plane moves (see pages 118–119).

Many pebbles, though, travel farther, carried in river water or washed by waves along a beach. As they travel, they collide with each other many times and, almost like being in a tumble mill that gradually smooths them down, this gives them increasingly rounded shapes. Beaches are the most effective at smoothing a pebble from the constant swash and backwash of water, and a pebble can lose half of its weight in just one season under such conditions. Depending on the kind of rock, pebbles can have different shapes. Rocks like slate, which split easily in one direction, typically form disc-shaped pebbles, while pebbles of rock such as granite are more spherical. A rock made up of these rounded pebbles is called a conglomerate.

Pebbles can travel even farther away from the coastlines where they are so effectively shaped. They can be dragged offshore during storms, and then swept along submarine canyons onto the deep-sea floor. Once there, there is little to make them travel farther: it is the end of the line for them. Buried under deep-sea muds, they will long survive as strata and as witnesses to such remarkable journeys.

▲ **Contrasting pebble histories**

(Left) Slate pebbles on a beach at Aberystwyth, Wales, have become rounded and smoothed by being continuously knocked against other pebbles as a result of wave action. (Right) These rock fragments, which were released onto a scree slope by erosion of the crags above, retain their original angular shape.

SAND AND SANDSTONES:
THE STORY OF A GRAIN OF SAND

In the making of each sand grain, there is a journey. And each journey – and each sand grain – is individual. However, some common patterns characterise the trillions of individual stories. One involves the mineral that most sand grains are made of: quartz.

A sand grain's journey often starts with the weathering and erosion of an igneous rock such as granite, where the dominant mineral – indeed, the most common mineral at Earth's surface – is feldspar, with quartz typically only making up the smaller part. But during the weathering – particularly in hot and humid climates – feldspar tends to break down chemically, its molecular structure transforming into the clay minerals that will be washed away to become mud. Quartz crystals within the granite are much more resistant to such chemical attack. As the rock around them decays, they are released as quartz sand grains. They are then carried by wind and water across the landscape and towards the sea, here and there piling up to form layers of quartz-rich sand.

Sand is not only quartz. Especially in dry landscapes, some of the feldspars survive long enough to form part of sand grains. They can often be identified by their opaqueness, in contrast to the glassy-looking quartz grains. On the beaches of volcanic islands, it is common to see black sands, made up of myriad tiny grains of eroded basalt rock, or sands full of bottle-green grains made of the mineral olivine, or even of mica flakes. Some kinds of limestone are made of sand-sized grains as well (see pages 92–95). Also, when examining with a microscope, an assortment of rare grains – garnet, zircon, tourmaline, apatite and other accessory minerals – may be seen. These give a distinct mineral 'fingerprint' to the sand; in ancient sandstones, geologists use these grains to work out what kind of landscape was eroded millions of years ago, to yield the sand grains that went into the sandstone.

Sand grains carried along rivers and beaches become rounded, just as pebbles do, but it is harder to smooth their edges off, as they are much lighter than pebbles and hence their collisions are not as energetic, especially when cushioned in water. The best rounded sand grains can therefore be found in sand deserts, where their collisions, driven by the fierce desert winds, are more violent. When studied up close, the grains are not only beautifully rounded, but have a frosted surface from the many impacts. Such telltale signs of desert sands can survive into ancient aeolian (wind-blown) sandstones, which help identify these spectacular strata.

▶ **Natural sandstone sculpture**

This sandstone on Animasola Island, Philippines, shows layers that reflect the different current speeds that sorted the sand grains into coarser and finer layers. Some of the currents were strong enough to drag large pebbles with them.

NATURAL WONDERS:
SAND RIPPLES AND SAND DUNES

When sediment particles are carried by currents of water or wind, they can form extraordinary, ever-changing and complicated structures that can then be 'frozen' in rock strata to give clues to conditions on Earth many millions of years ago. Some of the most striking structures – ripples and dunes – occur in sand. Today, they are most easily seen in layers of sand on a beach or tidal flat, and from the deep past they can be recognised in ancient sandstone strata.

These structures are a little mysterious. Physicists have puzzled over quite what function they have; they speak of the 'self-organising' properties of a sand layer when a current flows over it. Ripples made by water currents are the easiest to see. The sand organises into sets of small asymmetrical ridges that can be straight, or – when the current flows a little faster – break up into crescent-shaped forms. The water carries grains of sand up the long gentle slope of one side of the ridge, and the grains then avalanche down the steeper slope, building it up by forming sets of little inclined layers that face downstream. The sand is thus carried downstream in the form of these ripples, themselves migrating downstream.

▼ **Patterns of moving sand**
This windblown dune has a steep avalanche face on the right (and so it is migrating to the right, driven by wind from the left, through successive avalanche layers). The gentler 'stoss' surface is covered by wind ripples, which represent sand migration at a smaller scale.

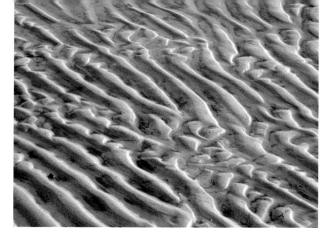

When these sand layers are petrified in ancient strata, one can sometimes see their top surface, just as one does on a beach or tidal flat. But more often, one sees the sections through them in the form of sets of these little inclined layers (as ripple cross-lamination) when the strata are cut or broken across. These are clues to the water current that formed them – not only to its speed, but to its direction, gleaned from the direction in which the inclined layers face.

Moving sand can also organise into larger structures: dunes. These do not form only by an increase in size of the ripples: they are a different class of structure (and can have ripples forming on top of them as well). However, they are similar in shape and behaviour to current ripples, and also produce advancing avalanche faces that produce inclined layers of sand that when fossilised in ancient strata produce cross-bedding (or cross-stratification). Dunes can be metres high, rather than centimetres, and can form from currents of flowing air as well as flowing water. In some respects, these are similar, despite the very different density difference between air and water (ripples can also form on wind-blown sand, but they are quite different in structure to current ripples in water, and form a different kind of layering, lacking cross-lamination).

When the current flows even faster – strong enough in shallow water to sweep a person off their feet – then the ripples and dunes are in effect washed flat, and horizontal layers of sand form. These often leave a telltale sign of this rapid current – tiny eddies swept into lines that produce characteristic parallel grooves in the sand. Ancient sandstones of this kind were often used for flagstones, because of the way they broke so nicely into flat slabs, and this characteristic grooving is often seen on them.

When paddling in the sea by a sandy beach, another kind of ripple may be seen, formed by the to-and-fro movement caused by the waves. Such wave ripples are symmetrical, unlike current ripples, and have sharp crests. In ancient strata, they are an unmistakable sign that these sediments were deposited in shallow water.

▲ **Modern and ancient sand ripples**
These ripples were sculpted in sand by flowing water: on the left, on a modern tidal flat on the German coast; on the right, 240 million years ago in what is now Utah.

The making of mud

Mud gets a bad press. If your name is mud among your friends, if you muddy the waters when trying to explain something, or if you sling mud when in political debate, your personal reputation will sink. This is a great injustice, if not personally then certainly to mud. For this substance is as close to a magic ingredient on a planet as it's possible to get, both as a springboard to life and as maintainer of that life – continuously, in the case of Earth, for at least three and a half billion years.

Like Earth, the planet Venus is a rocky planet; but it is a lifeless inferno, at least partly because there is no mud among those endless lava fields. Likewise, the Moon has some surface dust but, alas, no water to turn it into mud. Mars has just a little mud, formed billions of years ago, which is one of the reasons why space scientists suspect that life might once have emerged on that now-frozen planet. Earth, though, is smothered in mud, and is also home to a glorious range of life. This is no coincidence. So – what is this amazing stuff, mud?

The bare definition in geology, of sediment finer-grained than sand, tells only part of the story. The fine-grained dust on the Moon is just tiny chips of the Moon's primordial minerals, pulverised by meteorites. On Earth, and on Mars, most of mud's fine particles are of something new – the product of a molecular refashioning of the primordial minerals when water is added to the mix. The old minerals, born in incandescent magma, are torn apart on the colder, wetter surface, and new ones emerge. From such chemical weathering of primary rocks, the clay minerals can enter the scene.

Clay minerals are silicates, just like feldspar, pyroxenes and micas, but with some crucial differences. They most resemble the micas, having a sheet-like form on a microscopic scale, far too small to see

▼ Mud volcano crater, Buzau, Romania

There are thousands of mud volcanoes like this around the world. They are the surface expression of a vast subterranean plumbing system where liquid mud travels, driven by pressure changes. These muddy eruptions are particularly common where young, not-yet hardened, strata are accumulating, such as on large deltas.

HOW TO READ CLAY MINERALS

A scanning electron microscope (SEM) reveals the extraordinary variety of shapes and structures found in different clay minerals. Magnified thousands of times, they can appear like books with myriad pages, forests of hexagons, cigar-like rolls or tangled confetti.

● Dickite
● Kaolinite
● Chlorite
● Vermiculite

even with a hand lens or a standard light microscope. Under an electron microscope, however, the clay sheets reveal a fabulously baroque variety of shapes. These have another hidden story to read. All clay minerals have a huge surface area in relation to volume: a single gram can be spread out to cover hundreds of square metres. This vast, intricate surface area can act as the framework for more chemical reactions – and may well have been a kind of mineral scaffolding that helped to form and stabilise RNA and DNA at the dawn of life.

Life, once formed, needs a stable planet to continue existing – one that hasn't roasted (like Venus) or frozen (like Mars). Earth has avoided those fates largely, it is thought, because of chemical weathering. Carbon dioxide from the air gets involved in this process, and in so doing is slowly taken out of the atmosphere, preventing this greenhouse gas from building up to dangerously high levels. The end product, mud, is therefore a planetary life insurance. That's an idea worth bearing in mind the next time you dig the garden or walk across a ploughed field.

▲ **Jurassic Coast, Dorset, England**
Although mudrocks are the most common sedimentary rocks on Earth, this is not always obvious, since they are often soft-weathering and so tend not to form dramatic landscapes. But where they are cut into by rivers or the sea, as in these cliffs at Lyme Regis on England's south coast, one can begin to glimpse their abundance and their importance to our planet's construction.

MUDROCKS:
WITNESSES TO HISTORY

Mudrocks are the most common sedimentary rocks on Earth, though it usually does not appear so. They are generally softer and weaker rocks than sandstones and limestones, and so do not form such impressive cliffs and crags, but instead tend to occupy the less conspicuous low ground between them (see pages 40–41). Mudstones do not have such striking textures as those other sedimentary rocks, such as the cross-bedding that represents petrified dunes. But, once the key to interpreting them is gained, they are among the most eloquent witnesses to Earth's history.

Some of the clues are simple, such as colour. Red mudrocks, for example, are those in which the iron content (which does not have to be great) has been oxidised; typically, these represent conditions on land, especially in arid regions with little vegetation. In green mudstones, the iron is in a chemically reduced state, representing, for example, stagnant, waterlogged ground. Dark grey and black mudstones are carbon-rich due to a high content of decayed plant and animal matter, and laid down on poorly oxygenated sea or lake floors.

Some carbon-rich dark mudstones of this type represent local, biologically rich conditions in some part of a lake or sea. Some, though, represent more global events: times of a hot greenhouse world with sluggish ocean currents and an oxygen-starved sea floor. Dead plankton drifting down on that sea floor under those conditions did not so easily decay, but were buried in ocean floor muds. This was a way of sequestering their carbon in rock strata, rather than allowing it to seep back into the atmosphere as carbon dioxide. Over time, this process cooled the planet. Such black mudstones, therefore, are a part of Earth's thermostat.

▼ **Mudrock colour change**
The original red mudrock takes its colour from oxidised iron minerals. Once buried, oxygen-poor groundwaters circulated through joints (fractures) and along stratal surfaces, chemically reducing the iron to turn the mudrock green.

Under the conditions in which such black mudstones form, few animals live on the sea floor – there is simply not enough oxygen for them. However, the remains of animals and plants that fall in from the sunlit, oxygenated waters high above can be entombed in those muds, ultimately to become exquisitely preserved fossils – some even with petrified eyes, guts and skin. Some of these animal remains are preserved by iron sulphide as the mineral pyrite, which give these amazing fossils a beautiful golden sheen. Palaeontologists now avidly seek such exceptional fossil occurrences.

When the climate cools, ocean currents often speed up, and this helps bring oxygen back to the sea floor. Then, living animals such as worms and crustaceans return, and begin to crawl or burrow though these muds. The oxygen also speeds the decay of organic matter, so the muds are no longer as carbon-rich and become a paler grey in colour. The animals also churn the muds, so that the delicate layering is lost, replaced by the traces of burrows in the mudrock. This is a sign of Earth's thermometer adjusting to a different setting. Under these cooler conditions, the carbon is now beginning to return as carbon dioxide to the air to strengthen the greenhouse effect. Eventually, this will lead to warming again – so setting the stage for the black mudstones to return.

▲ **A carbon-rich mudrock**
The dark grey colour of this mudrock betrays a significant organic carbon content, the decayed remains of the tissues of organisms – many of them microscopic – that once lived on or above a muddy sea floor. The fossils represent the skeletons of some of them.

BELOW THE SURFACE:
OCEAN STRATA

The deep-sea floor includes the final resting places of much of the sediment that is eroded off the land. These sedimentary graveyards are enormous, and the easiest way to visit them is to look at ancient examples that have been heaved up onto land, to form the thick rock strata that make up much of the world's mountain belts. Looking at these strata shows the particular patterns that make them distinctive, and gives clues to just how the sediment ended its long journey.

The most widespread pattern is of a remarkably regular interlayering of sandstone and mudstone flat beds, often each just a few centimetres thick. The sandstone beds typically have very sharp bases, with the largest and coarsest grains at the bottom, gradually becoming finer upwards until the sand grades into mud – on top of which is the next sharp-based sand bed, and so on.

These strata have long puzzled geologists. The mystery was finally solved when it was realised that strata were formed by many individual strong currents, known as turbidity currents, carrying sediment a long way from shallow water into very deep water. These currents may be triggered by an earthquake or a big storm that destabilises masses of sediment (often billions of tons at a time), that then, under gravity, flow as a dense, turbulent current downslope, accelerating to express-train speed (such currents can rip apart stout submarine telephone cables). Turbidity currents can flow for hundreds or even thousands of kilometres, decelerating and carpeting the sea floor with a thick layer of sediment – a turbidite bed – that can entomb bottom-living animals over wide areas, the sand dropping out first, and then the mud settling more slowly on top. Turbidite strata overall can be kilometres thick; it has other telltale features as well, such as characteristic scours on the sea floor (flute marks), which are often preserved on the sandstone bed bases, as well as the marks of burrowing animals overwhelmed by this (for them) catastrophic event.

Enormous fans of such turbidite deposits drape the edges of the continents, and extend out some way onto the abyssal ocean plain. Beyond that, there are thin deep sea oozes, some of which are mostly made up of tiny skeletons of planktonic organisms that have drifted down from the sunlit waters far above. As we shall see, this creates a kind of rock called chalk, which marks one of Earth's most distinctive episodes.

▶ **Deep-sea strata**
These characteristically regular alternations of sandstone (pale) and mudstone (dark) are typical of 'turbidite' deposits that cover many deep-sea floors. Each sandstone–mudstone couplet results from the settling of a pulse of sediment from a turbidity current – the sand settles out first, and then the mud.

GRAVITY-POWERED SEDIMENTATION

The periodic triggering of gravity-powered pulses of sediment as turbidity currents is the main way in which sediment is transported out onto the deep-ocean floor.

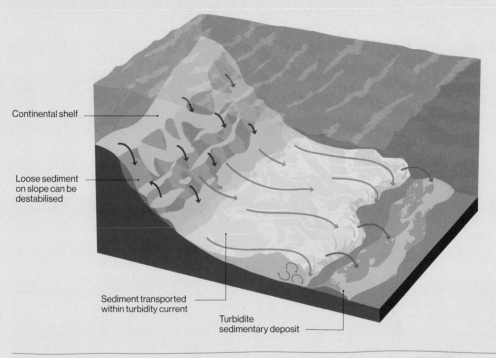

Continental shelf

Loose sediment on slope can be destabilised

Sediment transported within turbidity current

Turbidite sedimentary deposit

CHEMICAL AND BIOLOGICAL ROCKS:
FROM LIMESTONES TO PHOSPHATES

▼ Reef rock

This massive limestone on coastal Gotland, Sweden, is a mass of skeletons of corals, sponges, calcareous algae and other organisms that lived on a shallow sea floor some 430 million years ago.

Land surface weathering has led not only to the sediment washed into the sea, but also to all the dissolved chemicals that make the sea salty. Indeed, the sea is so salty that some of these chemicals can precipitate out again under the right conditions, to form strata of chemically formed rock, such as rock salt layers, when an arm of the sea dries up. Plants and animals can also extract dissolved components from the seawater to build their skeletons – over many generations, these skeletons build up on the sea floor, forming biologically made rocks. The most typical components here are calcium and carbonate, which make the mineral calcite (and related mineral, aragonite); these then accumulate to form limestone.

▲ An oolitic limestone

These are typical ooids magnified. They are a millimetre across or less, with a spherical shape and a layered internal structure. The five-rayed fossil is the part of the stem of a sea lily, a relative of starfish.

▲ A record of life

This 350-million-year-old limestone contains pale fossil corals. The subtle mottling in the rock represents burrowing by animals through the nutrient-rich limy muds of that ancient sea floor.

Such limestone formation can most easily be seen on a shell-rich beach and fossilised equivalents – the shelly limestones that represent ancient beaches and shallow sea floors – are a common kind of rock. In such shelly limestones, some of the fossils may be represented by the solid sea shells themselves (usually made by an animal out of the mineral calcite), while some are just empty shell-shaped spaces in the rock (these mostly represent shells that were made of aragonite, a harder form of calcium carbonate than calcite, but one that more easily dissolves underground, leaving a shell mould).

Another kind of biological limestone is that produced by a living reef. Today, these magnificent structures have been damaged by climate change and ocean acidification. However, their extraordinary living diversity, founded upon the framework provided by living corals, can still be seen. That underwater framework – the surface of which can be seen by scuba divers, say – extends deep underwater as complex three-dimensional accumulations of coral skeletons, among which are the fossilised remains of many other organisms. Such reef limestones can be very thick. In the 19th century, Charles Darwin first realised that coral atolls were built on volcanic islands that were slowly subsiding, while the coral organisms were always, generation by generation, growing upwards to remain at sea level. Such reef limestones can be a kilometre or more thick.

Today, reef limestones are built by corals, but in the geological past, other organisms, such as sponges and even different types of specialised shelly animals, could provide the framework. Biological reefs are delicate and easily killed off; they have appeared and disappeared through

▲ **Chert layers from half a billion years ago**

The tough chert layers of Oregon's Rainbow Rock are the hardened remains of oozes, made up of countless microscopic silica skeletons of radiolaria that fell to a sea floor in the Cambrian Period.

geological time. Humans activity appears to be causing another extinction event of this kind of ecosystem (and of the kind of rock it produces).

There are other kinds of limestone. Chalk, one of the world's most distinctive limestones, is found more or less worldwide. It is made up of countless skeletons (coccoliths) of planktonic single-celled algae that rained onto a sea floor in the Cretaceous Period 80 million years ago, when the world was very hot and sea levels were very high.

Another distinctive kind is chemically formed: an oolitic limestone. This is made up of ooids: sand-grain-sized spheres. Ooids form in warm, shallow, wave-swept seas (such as around the Bahamas today), a bit like the way hailstones form in the atmosphere, by accumulating microscopic layers of calcium carbonate around a central core. Oolitic limestone makes good building stones, and so is commonly found on town and city buildings, where its intricate detail can be conveniently examined with a hand lens.

While limestones are the most abundant kind of chemical-biological rock, other varieties are both distinctive and important – not least to help keep humanity alive!

Silica-based sedimentary rocks exist other than those made of quartz grains. One kind is made of the intricate and beautiful microscopic skeletons of silica-secreting plankton such as radiolarians (single-celled amoeba-like animals) and diatoms (single-celled algae). These can be found either where there is an abundant supply of dissolved silica (as in volcanic lakes, where diatoms can proliferate), or where other kinds of sediment are excluded, such as on the very deep ocean floor. Here, far from land-sourced sediment and where calcite skeletons dissolve, layers of radiolarian skeletons slowly build up as oozes.

When these strata first form, the layers are loose and powdery (and the complex skeleton structures make them very useful for industrial filters, as with 'diatom earths'). However, when they are buried, the heat and pressure make the skeletons (which are of opaline silica, containing water in its structure) recrystallise into a solid, very tough rock called chert, in which the mineral structure is now quartz. Such chert layers are often found in mountain belts, where deep ocean strata have been crumpled and pushed up to form land.

One common form of chert is flint. This is most typically associated with chalk, in which it forms layers of irregular nodules, some small and potato-shaped, others larger with more complex shapes. These flint nodules formed underground, when fossils with silica skeletons buried in the chalk layers (commonly glass sponges) dissolved, and the silica was re-precipitated within the chalk as flint nodules. Flint is extremely tough and often goes on to form abundant pebbles anywhere chalk strata are eroding.

A mineral very close to humans is apatite, chemically known as calcium phosphate, as it makes up our bones and teeth. It is also the stuff of the bones of dinosaurs, mammoths and other vertebrates (it makes up some kinds of shell, too). Sometimes layers of bones, scales and teeth can be swept together by marine or river currents, which winnow away lighter sediment grains to concentrate these fossils as a bone bed. Such rare strata can be spectacular, lending an awe-inspiring window into the animal communities of the past.

Another form of fossil phosphate derives from fossilised animal dung known as coprolites. Coprolites can provide important clues about the diet of prehistoric animals. Phosphate-rich strata, including coprolite-rich beds, are crucial to humans, because they are the main source of phosphate fertiliser, which is vital to agriculture. Commercially sized deposits are not common, not found everywhere in the world, and are non-renewable. It is thought that the world may be approaching 'peak phosphate', so this kind of rock will become even more important in the future.

▼ **Coprolites**
These irregular phosphorus-rich masses are coprolites – the fossilised remains of animal dung. They can be found in land-derived strata, from animals such as dinosaurs, and also in marine strata, from sharks and marine reptiles.

TELLING THE TIME:
FOSSILS

Earth's rock strata cover a period of more than three and a half billion years. Navigating the different ages of Earth's history can be done in different ways. For example, one can exploit the natural radioactivity of a rock to work out an age in millions of years. With sedimentary rocks, the best and easiest way is to use fossils. While it cannot give an age in millions of years, it draws on the history of Earth's constantly changing life forms to allow the placing of rock strata in relative order: that is, to say which stratum is older and which is younger.

For the first three billion years of strata of Precambrian times, fossils provide only a very crude measure of time. Indeed, in the early days of geology, these rocks were thought to lack fossils completely. But later study showed that they included microbially layered rocks called stromatolites, and sometimes the microscopic remains of the microbes themselves. These, though, are rare and show only gradual change through that long time span.

Then came a biological revolution. A little more than half a billion years ago, and within a (geologically) short span of time (just 30 million years), the whole range of main animal groups that we know today evolved: molluscs, arthropods, worms and others (including the first fish, albeit these were very rare then). This explosion of life marked the beginning of the Cambrian Period, and of complex life as we know it today. The subsequent evolution of that life, preserved as abundant fossils, has now been worked out in much detail (though there is still

▼ **A very imprecise fossil clock**

Stromatolites – fossilised layered microbial structures – dominated the first three billion years of Earth's fossil record. Even across this immense time span their preserved patterns show few obvious changes. Their geometry and appearance in the rock, though, can be quite variable, as shown by the two specimens below.

◀ **Fossil time-markers**
These fossils are trilobites, typical fossils of the Palaeozoic Era. Within the nearly 300 million years of this geological period, many thousands of species evolved. By recognising these individual species, palaeontologists can identify precise 'time slices' within the rocks that may represent less than a million years.

much to learn). It provides a wonderful chronometer of this most recent phase of Earth history, which we call the Phanerozoic Eon (ongoing today).

For the specialist, this fossil chronometer can provide a high degree of precision, recognising units of time of less than a million years, hundreds of millions of years ago, based on the appearance and extinction of individual fossil species. But one can use fossils more generally too, to recognise strata of the three great eras of Phanerozoic time: the Palaeozoic, Mesozoic and Cenozoic.

Fossils that are clear markers of the Palaeozoic Era include the iconic trilobites, marine arthropods that ranged right through this era (but which were more abundant in its early part, with thousands of different species); their three-lobed carapace is characteristic and distinctive. The planktonic graptolites – complex animal colonies that look like pencil drawings when fossilised – are also typical.

The Mesozoic is most known for being the time when dinosaurs lived – but dinosaur bones are generally rare, and so these magnificent fossils are not the best practical guide to time-telling. The small fry are the much more effective time-markers, and among these the ammonites and belemnites – both being relatives of squid and octopuses – are common and distinctive.

The Cenozoic Era – in which we still live – is when 'modern' ecosystems developed. On land, the mammals flourished but, as with dinosaurs, their remains are rare. Fossils of mollusc bivalves and snails are much more common. Another set of unmistakable Cenozoic fossils includes the nummulitids: known as 'giant microfossils', they were dime-sized with many-chambered discs. These could be abundant; the Egyptian pyramids are built of nummulitic limestones.

▲ **Belemnites**
These fossils are common in rocks of the Mesozoic Era. Superficially very similar, palaeontological specialists can identify subtly different kinds, and use them as time markers in strata of this era.

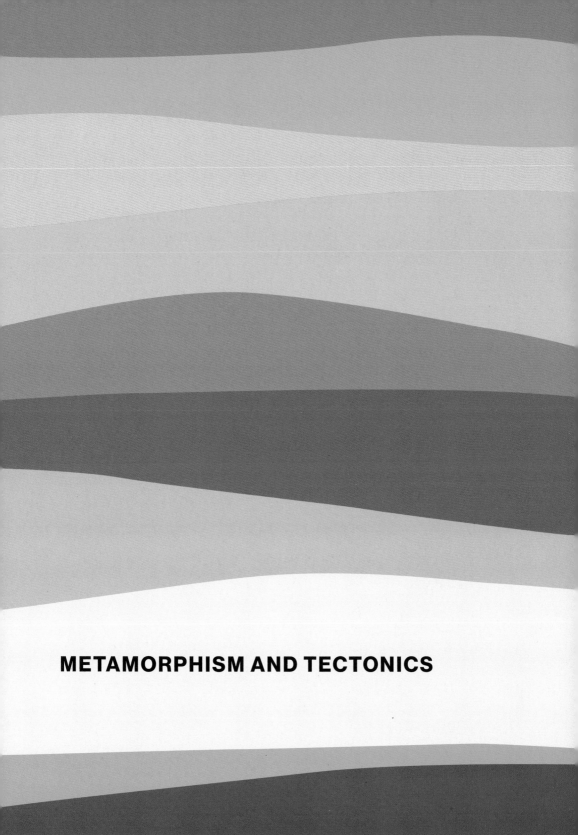

METAMORPHISM AND TECTONICS

THE RISE OF STRATA:
MARINE DEPOSITS AT ALTITUDE

The occurrence of petrified marine shells high up on mountainsides intrigued and puzzled early scientists, including Leonardo da Vinci in the 15th and 16th centuries. Leonardo saw that shells in the mountainside strata around Florence, Italy were in orderly layers and often located where they had lived, not scattered from some violent deluge. He deduced that the 'bottom of the sea was raised', so that the Arno River could now cut through it. Today, we know such occurrences extend to the top of the highest mountains of all, as fossil seashells, trilobites and sea lilies, half a billion years old, have been found in the strata forming the very summit of Mount Everest, 8 kilometres (5 miles) above sea level.

▼ **Lifted by the tectonic escalator**
Marine fossils have been found at the very summit of Mount Everest. The tilted and crumpled strata that can be seen on such mountains are a clue to the tectonic forces that have lifted these parts of the land so high.

Leonardo's keen intuition was typically insightful. Modern science shows that Earth's crust can indeed rise and fall (and sea levels can rise and fall independently of the movement of the crust). Early geologists considered many ideas as to how this might have worked. The late 18th-century Scottish geologist James Hutton suggested that Earth acted as a gigantic heat engine. This is essentially the case, with Earth's tectonics governed by the need for it to release its inner heat. This happens when the crust moves sideways, in plate tectonics (see pages 28–29), with hot magma welling out to cool and form the spreading ocean-floor crust.

The local rise and fall of the land occurs in response to the movement of tectonic plates, seen in mountain belts such as the Himalayas and Andes. Here, deposits once on the deep ocean floor can be scraped off a descending tectonic plate onto the edge of a continent, raised high as the rocks are crumpled to form mountains. Many belts include large amounts of deep-sea rocks such as turbidites (see pages 90–91) as a result. In the process, rocks can be jammed down into Earth to 16 kilometres (10 miles) deep, before being rapidly squeezed back to the surface. Such rocks include blueschists, containing high-pressure minerals with a distinctive blue tinge.

Away from intense crumple zones, where rocks are altered greatly by heat and pressure, crust rising and falling can be slower and gentler – much less damaging to the fabric of the strata and the fossils they contain.

In Cretaceous times, for example, the centre of North America was occupied by a wide, shallow sea. The crust beneath this inland sea subsided slowly, so that layers of sediment accumulated across the area. Later, they slowly rose to bring these strata towards the surface again. Strata like these have neither been strongly crumpled, nor greatly altered by heat and pressure – just compacted under their own weight, the deepest parts gently simmered by Earth's heat.

▲ **Gentle tectonic fall and rise**

The classic flat-lying strata of Arizona's Grand Canyon are rocks that have been gently buried, without tectonic crumpling, and then raised to the surface again, to be eroded now by the Colorado River.

TECTONICALLY COMPRESSED MUD:
THE MAKING OF A SLATE

Hike or drive across a mountain belt today, and you will be crossing rocks from different parts of its ancient interior, now exhumed from great depths as the earlier generations of mountains that once lay above were eroded away. Many of these exhumed rocks are metamorphic, altered by heat and pressure from their igneous or sedimentary origins.

In the outer parts of a mountain belt, the original rocks have not changed beyond recognition, but nevertheless have transformed into a different kind of rock. Slate is a fine example of this. Originally mud on a sea floor, it was buried and compacted by the weight of strata above to form a mudrock. Then – if caught between colliding tectonic plates in a growing mountain chain – it was unmercifully squeezed, to be crumpled into giant folds.

If this crumpling takes place more than 8 kilometres (5 miles) below ground, where the temperature reaches 200°C (392°F) or more, the texture of the rock slowly undergoes a fundamental change. The tiny flake-like clay minerals in the rock begin to transform: still microscopic, they change their alignment to lie flat against this new vice-like pressure from the side. The rock now no longer splits along its original layers, but usually at a high angle to them, along these newly formed mineral planes. It has become a slate.

If the original mud was pure enough, the slate can be split into thin roofing tiles by a skilled artisan. Look closely at the surface, and faint traces of the original lamination (layering) or animal burrows can be seen.

Most mudrocks in nature, though, are not of such purity, but contain silty or sandy layers, unsuitable for tiles – or billiard tables! However, the rough, sandy layers contrast with the smooth splitting planes of the slate to beautifully illustrate the history of the rock. Lacking flaky minerals, they do not split so cleanly – but the quartz grains are tightly recrystallised to form a tough quartzite.

Other changes can take place as slate forms. If the mud contains iron sulphides, sometimes these recrystallise as large cubic crystals of pyrite (fool's gold) within the slate. If the rock is fresh, these keep their metallic golden colour. If the rock has been weathered, the pyrite can decay to leave conspicuous cubic holes in the rock. These can look a little surreal – but are perfectly natural.

▶ **Slate: a mudstone refashioned**
This rock splits along myriad parallel planes of slatey cleavage imposed upon the rock when it was tectonically compressed. Faint traces of the original stratification can be seen at a high angle to these cleavage planes, which pick out the shape of a tectonic fold.

SLATE FORMATION

Lateral compression of
buried mudrock strata first
crumples them into folds,
and then reshapes the flake-
like clay minerals, so they
regrow perpendicular
to the pressure, which
gives the splitting direction
of a slate.

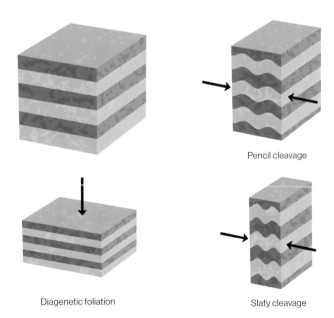

Pencil cleavage

Diagenetic foliation

Slaty cleavage

THE HEART OF A MOUNTAIN BELT:
SCHISTS, GNEISSES AND MIGMATITES

Consider where rocks are being crumpled a little deeper
in the mountain belt than where slates form – say a few
kilometres deeper and a hundred degrees hotter. There, the
flaky minerals that were once clay now begin to grow larger,
and turn into shiny mica crystals. The rock has turned from a
dull slate into a shiny phyllite – and this change is a gateway
to further transformation, as the heat and pressure increase.

Go deeper beneath a growing mountain belt – about 10 to 11
kilometres (6 to 7 miles) down at temperatures exceeding 400°C
(752°F) – and the rocks are even more strongly metamorphosed,
almost unrecognisable from what they once were. Within a former
mudrock, then a slate, then a phyllite, mica crystals can grow larger,
now forming parallel layers between layers of recrystallised quartz to
become a gleaming mica schist. In some, marble-sized, blood-red garnet
crystals have grown. At this stage, nearly all fossils within the rock have
been obliterated.

▼ **A telltale shine**
This phyllite pebble has
a characteristic shiny
surface, produced by tiny
mica crystals. The little
tectonic crumples are
also characteristic, while
the small brown patches
are weathered remains
of pyrite crystals.

▲ **Twin crystals**

This is a type of twin crystals of staurolite (other staurolite crystals can intersect at right angles, to form perfect crosses), which form at high temperatures and pressures in metamorphism.

▲ **Newly grown garnets**

The characteristic red crystals among large shiny micas and recrystallised quartz give a garnet-mica schist its distinctive texture among metamorphic rocks.

Where temperatures and pressures become higher still, further mineral transformations can take place. The garnet disappears, to be replaced by crystals of higher-temperature metamorphic minerals such as staurolite, which can form distinctive 'twinned' crystals in the shape of a cross. Other rocks at these temperatures and pressures metamorphose in different ways, depending on their original chemistry. A limestone recrystallises to form marble, a processes that destroys fossils it once contained. A basalt becomes an amphibolite, full of the dark mineral amphibole.

Journey even deeper – about 16 kilometres (10 miles) down at temperatures exceeding 500°C (932°F) – and the extreme metamorphism continues. Now, in what was once a shiny mica schist, most of the micas break down and transform into feldspars, which, with quartz, make up most of the rock. This rock has become a gneiss, still with a roughly streaky, banded appearance, as it is still being squeezed and sheared between colliding tectonic plates.

During this transformation, the rock remains solid. But in the heart of the mountain belt, at maybe 25 kilometres (about 15 miles) down and near 800°C (1,472°F), conditions can become yet more infernal. Here, a gneiss can begin to melt and produce patches of a magma that is like granite in its composition. This granitic magma can form sheets between layers of still unmelted gneiss – now a mixed rock, a migmatite.

When temperatures reach maximum heat, the melting is finished, producing a magma that will eventually cool to form granite. The rock cycle is complete. First, an ancient granite has been weathered to form mud. Next, this has formed a mudstone, and finally – via metamorphism to slate, schist and gneiss, and melting to a magma – it becomes granite again.

CRACKING OPEN: WHERE TECTONIC PLATES PULL APART

To form an overall view of where tectonic plates pull apart, a globe is useful in showing the shape and location of the oceans. However, a map of Earth that omits ocean water is even more helpful. On this type of a map, the sharp difference between the youthful crust of the oceans and ancient continents is clear. It also shows something that was only recently discovered: in the middle of the ocean basins are mountain chains – not ones formed by compression, like those on land, but those formed due to the stretching of the crust. Where this crust is cracking open, magma continually wells up to form new ocean-floor rock (which is hot, so less dense; hence, it rises higher to form undersea mountain ranges). Known as mid-ocean ridges, these are part of the mechanism of plate tectonics (see pages 28–29).

WHERE OCEAN CRUST FORMS

This map shows the courses of the mid-ocean ridges, which almost always lie submerged deep below the sea's surface.

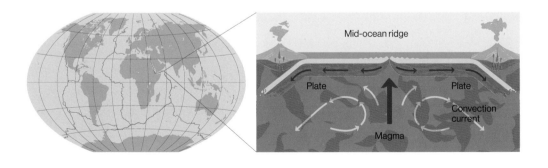

▲ Where Iceland is breaking apart

Magma rises to the surface along Iceland's rift zone, to form new ocean crust (albeit here above sea level).

Most of this activity takes place underwater, so is hard to see. But enough is known to say the new rock (and therefore, all the ocean crust) is almost all basalt – basalt that has an extraordinarily uniform composition across Earth. It is subtly chemically different from other types of basalt – like the types that build up ocean islands like Hawai'i, where magma punches up through the ocean crust rather than welling up at its growing edge. Some of this magma reaches the sea floor as eruptions of pillow lava; some is injected into cracks in the crust as swarms of dykes (see pages 52–53).

There is one place on Earth where part of a major mid-ocean ridge rises up above sea level, showing how a tectonic plate is pulled apart. This is Iceland, which has been pushed upwards as a column (or 'plume') of mantle rock slowly rises beneath it, like a gigantic subterranean fountain. On Iceland, one can see fissures where the ground has been stretching apart (by 70 metres, or 230 feet, in the last 10,000 years), some as spectacular canyons, while the frequently erupting volcanoes are part of the magma supply that enables Iceland's crust to grow.

Other stages in the growth of ocean floor can be seen elsewhere. The Red Sea is a baby ocean, about 320 kilometres (200 miles) across. Its growth – by a centimetre a year – is causing Arabia and Africa to move ever farther apart. An even earlier stage in the process is seen in Africa's volcano-lined Great Rift Valley, which marks the beginning of the cracking up of the African tectonic plate. If this crustal stretching continues, Somalia will separate from the rest of Africa in about 10 million years.

EPIC IMPACT: WHERE TECTONIC PLATES COLLIDE

The results of tectonic plate collisions make up some of the most spectacular parts of Earth's crust. They have resulted in the highest places on Earth, among the peaks of the world's mountain ranges – and also the lowest, where the sea floor plunges to great depths. They display an astounding range of rocks – so great, that it took geologists some time to work out their fundamental patterns and meaning to our planet.

▲ Collision landscape
The world's high mountain ranges are areas where Earth's crust has been crumpled and thickened by collision between tectonic plates. Where the plates are still colliding, the mountains continue to rise, as shown here in Tierra del Fuego, South America.

One type of collision is where one oceanic plate collides with another, such as just west of the Marianas Islands, where ocean crust of the Pacific is moving towards and sliding below that of the Philippines. There is a great crescent-shaped arc where this Pacific crust flexes downwards to sink into Earth's mantle, marking an ocean trench, the Marianas Trench, which includes the deepest point on Earth: the Challenger Deep, 11 kilometres (7 miles) below sea level, about twice the depth of the normal ocean floor. It is this deep partly because the sinking ocean crust is old and dense, and partly because the Marianas Islands, which run parallel with the trench, are small and cannot supply much sediment to fill the trench. The Marianas Islands themselves are a result of this process of tectonic subduction, made of new volcanic rocks formed from magma rising from where the Pacific plate is descending. These volcanic rocks are more silica-rich, and therefore less dense, than the basalt of the surrounding ocean crust, so rise higher above the ocean floor. These islands are, in effect, a piece of new continent in the making.

THE DESTRUCTION OF OCEAN CRUST

This map shows the location of the Mariana Trench, which includes the deepest point on the sea floor. Here, Pacific Ocean crust is subducted down into the mantle.

Another kind of collision is between an ocean plate and a continent. The classic example is where the Pacific ocean crust is moving towards, and subducting beneath, the continental crust of South America. The result is the Andes mountain chain. Here, the mountains rise so high partly because of the crumpling crust of western South America by the forces exerted by the colliding Pacific ocean plate, and partly by the growth of large explosive volcanoes such as Chimborazo and Cotapaxi – part of the Pacific 'Ring of Fire' of volcanoes, earthquakes and tsunami.

There are also collisions between continents. These start off as collisions between a continent and an ocean plate carrying another continent. When the ocean plate between them is destroyed by tectonic subduction, the two continents come into contact. This is what happened when India, once part of an ancient supercontinent called Gondwana, broke free of it and drifted northwards towards Asia, the ocean between them destroyed in the process. About 50 million years ago, these two continents collided. They are still colliding like in a slow-motion video: neither can be subducted, as both are too light.

IN OPPOSITE DIRECTIONS:
WHERE TECTONIC PLATES SLIDE

Tectonic plates, as well as splitting apart and colliding with each other, can also move past each other. Even when they split apart and collide, there is usually some sideways motion involved, and so they approach or move away from each obliquely. But there are some places where tectonic plates slide directly past each other. The best-known example is in western North America, where the San Andreas Fault marks the boundary between the North American plate (moving south) and the Pacific ocean plate (moving north). Even this is not a simple 'slide-past' motion, as the two plates press upon each other, too, helping raise mountains in its vicinity. It is not a simple fracture in the crust, but includes different strands arranged in a complex pattern.

The movement along the San Andreas Fault is inexorable, though eventually, millions of years from now, Los Angeles will slide past San Francisco and carry on moving towards the Aleutian region. The movement of these two huge crustal blocks past each other is notoriously expressed as sudden, sharp movements that, without warning, can generate powerful earthquakes such as the 1906

▼ **Crustal fracture**
Part of the San Andreas Fault system, along which the Pacific and North American plates are sliding past each other. The rocks around the fault zones have been intensely shattered by the forces released during multiple earthquakes.

earthquake that devastated San Francisco. Such earthquakes are inevitable in the future, given the forces involved. So far, no means have been found of forecasting precisely when and where they might take place. Earthquakes, when they occur, still surprise us.

Part of the movement of the rocks along the fault, though, is more gentle, a kind of slow, almost imperceptible, non-destructive creep. This kind of gentle movement is helped by what has happened to the rocks near the fault. The many wrenching movements of earthquakes have shattered and sheared them in a broad zone along the fault. This shattered zone is permeable to subterranean fluids and these, slowly percolating through the fault zone, have chemically altered the rocks, converting some of their components into slippery minerals such as clays. There is therefore a broad, weakened zone around this major fault, which can slowly and quietly yield to the relentless plate tectonic forces, as well as, sometimes, suddenly snapping in a catastrophic earthquake.

The action of this plate boundary over millions of years has not only affected the rocks underground: it has had a huge impact on the landscape, too. To be more precise, it has slid landscapes past each other, so that the pattern of hills and valleys often changes dramatically on crossing the fault. This has even affected geologically recent landscapes, so that the upstream and downstream ends of rivers flowing across the fault have been dislocated. It is one of the most striking examples of Earth's dynamic landscape.

THE HEAT EFFECT:
BAKING BY MAGMA

The heat of magma, which can exceed 1,000°C (1,832°F), affects rocks in various ways, whether at Earth's surface or deep within Earth's crust.

An advancing basalt lava can bake the soil beneath, and burn off organic matter within it. The soil, reddened, compacted and hardened in this way, can even sometimes develop vertical column joints, smaller versions of the spectacular columns that can form in the basalt itself.

A magma coming into contact with rock can similarly bake and harden it. Or, if the rock is friable like a soft, wet sandstone or mudstone, the intense heat can flash the water to steam, blasting the rock apart and mixing it with shreds of magma. The resulting mixture can form thick layers as the advancing magma 'steam-blasts' its way into the rock. Once cooled and solidified, the resulting rock is called a peperite (because the geologist who named it thought it looked a bit like coarsely ground pepper).

Deeper underground, other kinds of rocks form next to the large bodies of magma that will eventually solidify to form granite or gabbro plutons. The cooling here takes place slowly, over many thousands of years, allowing different kinds of new minerals to form, in contact metamorphism.

The minerals that form here depend on what kinds of rocks are being heated. If a granite magma comes up against layers of dark, carbon-rich mudstone, then the very distinctive aluminium silicate mineral chiastolite can grow in the rock. Chiastolite forms white crystals about the size and shape of a matchstick that, when broken across,

show a cross-shaped dark area, made of the carbon particles that were concentrated inside the crystal as it grew. If heated for longer, the entire mudrock can be recrystallised, and produce a very dense, solid rock called a hornfels.

If a granite intrusion comes up against a limestone, a skarn can form, with minerals that represent the combined chemistry of the limestone and the granite. Valuable metals such as tin, copper, gold and nickel, can also be concentrated in skarns, by the heated underground fluids circulating around the cooling granite – and so these are much sought after by prospecting geologists.

CONTACT METAMORPHIC TRANSITIONS

The kind of rock that results from thermal baking depends on the temperature and the kind of original rock that comes into contact with magma, as well as on the size and temperature of the magma body.

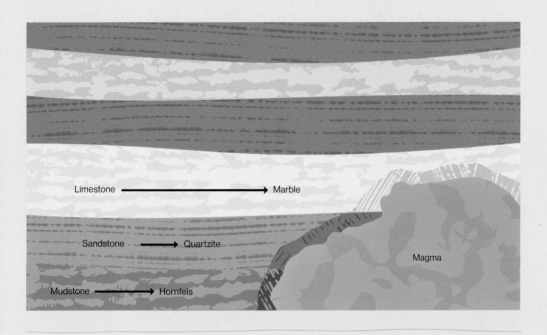

Limestone ⟶ Marble

Sandstone ⟶ Quartzite

Mudstone ⟶ Hornfels

Magma

◀ **Lava baking**

A reddish baked soil, sandwiched between two basalt lavas, photographed along Road I-17 near Flagstaff, Arizona. The temperature of the overlying lava was about 1,100°C (2,012°F) when it buried the soil, and this high temperature dried out the soil and caused the formation of prominent columnar joints.

▶ **When magma meets wet rock**

When magma was injected into water-soaked rocks, more than two billion years ago in what is now Ontario, Canada, the resulting steam blasted it apart into fragments, forming a rock called peperite.

HOT WATER UNDERGROUND:
THE MAKING OF MINERAL VEINS

Earth contains a marvellous storehouse of ore minerals –
probably richer and more diverse than on any other planet in
the solar system. The recipe for these riches includes a wide
variety of rocks, Earth's internal heat – including the local
heat engines of magma chambers and volcanoes – and an
abundance of water coursing through the hot rocks at depth.

The most common and most eye-catching symbols of this
phenomenon are the white mineral veins – typically made up of quartz
– that can be seen zig-zagging across many cliffs and crags in rocky
terrain. In a nutshell, these form because of the stewing up of rocks at
depth – and with them are different kinds of planetary kitchen.

One of the simplest is a mass of muddy strata buried by a few
kilometres' thickness of other strata, especially if this is also squeezed
during a mountain-building event and converted to slate (see pages
102–103). The squeezing wrings out these rocks, driving out fluids with
dissolved minerals, including dissolved silica. These fluids work their way
upwards into regions of the rock that are cooler and at lower pressures,
then find their way into fissures and fractures. Here, silica crystallises to
form quartz, which is often characteristically milky white rather than
clear and transparent, due to myriad gas bubbles trapped in the growing
crystals. The fluids are prone to boiling due to leaks and sudden pressure
drops, and many of the minerals form in such boiling episodes.

As well as silica, the hot, wrung-out fluids can bring with them
metals such as barium, copper, lead, zinc, tin and gold, for the original
muds contained low concentrations of such elements. In some veins,
these can become greatly concentrated as crystals and layers of metallic
minerals such as barytes, chalcopyrite, galena and zincblende, while
gold simply forms crystals and nuggets of the metal itself. These veins,
therefore, contain riches that humans have been aware of since they
started using metals. But they are not easy to access.

Volcanoes and the underground magma chambers that supply
them represent another type of 'stewing'. The magma not only brings
its own chemical cocktail, but a source of underground heat that can
persist for many thousands of years before the whole system cools.
This heat causes the mineral-laden fluids to rise upwards towards the
surface to form veins, then emerge at the surface as hot springs and
geysers. And, as these hot fluids rise, cold rain-fed fluids are drawn
in from the surface regions towards the subterranean regions of the
volcano – to become heated, rise and carry minerals in turn.

▶ **Mineral veins reach
the surface**

Where hot, mineral-
laden waters from deep
underground reach the
ground surface, they
can form spectacular
hot springs, like these
examples from Hverir,
Iceland. Minerals can
be deposited in this
setting, too.

TECTONIC FORCES:
CRUMPLING ROCKS

One of the most profound effects of the forces that shape Earth's crust is the way that solid rocks can be folded, sometimes on a huge scale. These spectacular structures are omnipresent in mountain belts, where they are produced by the compressional forces unleashed by colliding tectonic plates.

▼ Crumpled layers
The strata here features a strongly folded unit between two undisturbed sedimentary layers. This shows that the folded strata must have been deformed on the floor of a sea or lake, with sedimentary layers laid down on top of them.

Rock strata can be folded in other ways, though, that do not involve enormous tectonic forces. Gravity can act on near-surface soft, sticky strata, to make them slide down a slope and crumple like a tablecloth. This process can produce eye-catching convolute bedding. Common in some kinds of strata, it can be mistaken for tectonic folding, but is usually distinguished on closer inspection. One telltale sign is where strata just above and below the crumpled layer remain even and undisturbed.

Where tectonic compression is very slight, the resulting rock folds can be large and gentle – to such an extent that it is difficult to see them as a whole. In one place, the strata look gently tilted in one

direction, while perhaps a kilometre away, gently tilted in the other. One can walk across the countryside, systematically measuring these tilts and plotting them on a map, to demonstrate such folds.

As the compression increases, the rock folds become tighter and more clearly visible. Where they form deep underground, mudstones can turn into slates (see pages 102–103). These can take different shapes: some are asymmetrical, with one side steeper than another. The steeper side (or 'limb') can show where the main tectonic 'push' came from.

With even greater pressure, the two limbs of the fault can become parallel, and the whole stack of folded rock may slowly topple over due to gravity and/or shearing forces. The result can strongly resemble simple, undisturbed beds of strata – only some of the layers are the right way up and others are upside down! Called recumbent folds, these are common in mountain ranges. To recognise them and work out their structure, geologists pay great attention to 'way up' structures, such as fossilised burrows or scours, which indicate the original top and bottom of a bed of rock.

Mountain belts form over tens of millions of years. This is long enough for rocks to be folded, then refolded more than once, as the patterns of tectonic forces change. The resulting tangles of strata make fine puzzles for geologists, who do their best to reconstruct the forces that formed the mountain belt.

Rocks can be folded in other ways, too. They can be deformed by a large ascending mass of granite magma, or by ascending plumes (diapirs) of rock salt.

▲ **Tectonic rock folds**
These are typical tectonic folds, formed when the crust that these strata were lying on was compressed and shortened. These kinds of folds are common on convergent tectonic plate margins. In some places the softer mudrock (the dark layers) has been squeezed into the hinges of the folds, while the pale sandstone layers, being harder, retain their shape.

PRONE TO FRACTURE:
BREAKING ROCKS AND FAULT ZONES

When pressure is applied to rocks, they can bend and form folds. However, they can also break, and the broken surfaces can move past each other along tectonic faults. What makes rocks break rather than bend? The first factor is the type of rock involved. Some types are more brittle and naturally tend to fracture, such as hard sandstones, limestones and granites. Others are more ductile and bendable, like mudstones and rock salt beds. The second factor is time: if pressure is slowly applied, then the rocks are more prone to bend, but if applied more quickly, breaking is more likely, as seen when manipulating modelling clay.

In areas where rocks are deformed, both faults and folds can form, and they typically occur together. The fold movement is gradual, but much fault movement takes places abruptly, when the friction on the fault plane is overcome so that the rocks snap into a new position. The resulting shock causes earthquakes: these vary in scale from catastrophic to barely perceptible, following fault movements from substantial to tiny. Faults often form where Earth's crust is being stretched, and here, one slab of rock typically slides down along a steep fault plane in the rock. This is termed a 'normal fault', the most common type of fault. Some well-known examples occur as facing pairs that define a rift valley, such as the Great Rift Valley in

Africa (see page 107), within which a large, long slab of crust has dropped down.

If the rocks are being compressed rather than stretched, one slab of rock can be pushed up past another, along such a steeply inclined fault plane, in a reversed fault. Where the compression is intense, such a slab can be pushed up along the ramp-like, gently inclined fault plane of a thrust fault. Thrust faults are common in mountain belts, and typically bring masses of older rocks to lie on top of younger rocks.

In another kind of fault, rock masses slide past each other sideways along a fault plane. A gigantic example is the San Andreas Fault, which marks one of Earth's major tectonic plate boundaries (see pages 110–111).

What about the fault plane itself? In huge faults that have moved many times, such as the San Andreas, the fault plane can be a wide crush zone of rock. But in smaller faults, it is sometimes possible to see the distinct rock plane along which movement has taken place. This often shows grooving and tiny rock steps, indicating which way the rocks have moved. Many fault planes are mineralised, too, due to crystallisation from the subterranean fluids that have flowed along them.

TECTONIC FAULT GEOMETRIES

Faults are classified on the basis of how the crustal masses slide past each other along the fault plane.

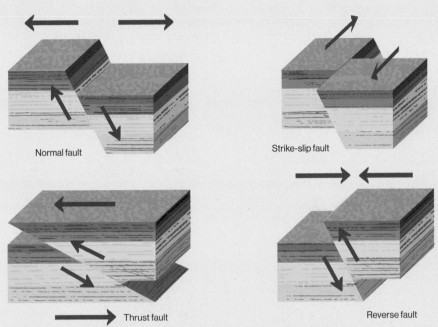

Normal fault

Strike-slip fault

Thrust fault

Reverse fault

◀ **Minor fault**
The strata here have been offset by movement along a minor fault plane. The rocks on the left slid down the slope of the fault plane relative to those on the right (best seen by the offset of the distinctive pale layer), showing that it is a normal fault, caused by the strata being stretched. Such small tectonic faults are extremely common.

TOPOGRAPHICAL CLUES:
READING TECTONIC LANDSCAPES

One can go to rock crags and pore over the detail, to see how the rocks have been affected by earth movements. But one can also stand back and look at the landscape, for clues to how Earth's crust has been moving.

▼ Death Valley's crustal movement

The mountainous regions either side of the valley are rising, and are constantly eroded. The valley itself is tectonically subsiding, and being further filled with sediment derived from the mountains.

One simple guide to movement is the general shape of the land. Where one can see steep craggy mountains like the Rockies, that is usually a sign that the entire landscape is still rising – and indeed that mountain-building processes are taking place far below our feet. On the other hand, where the land is low-lying and flat, it commonly means the landscape is stable or is subsiding and being continually buried beneath new sedimentary layers. This is especially the case where there are thick accumulations of recent strata near the surface, such as coastal areas of the southeast USA, or of the Netherlands, or Bangladesh.

The thick brown
sandstone bed the
geologists are standing
on suddenly disappears in
the centre of the image – it
has been displaced along
a near-vertical fault plane
that cuts across it. One of
the tasks of a geologist is
finding this sandstone bed
on the other side of the
fault, to work out how big
the displacement is and
in what direction.

Even in generally rising mountainous regions, the complex patterns of
forces may mean that some areas are subsiding, while the ground just
next to them is rising. An example is Death Valley, which is a slab of
crust that is moving downwards – a graben – between mountains on
either side that are rising. The sharp contrast between the flat floor and
the sharply rising mountains on either side marks the positions of the
tectonic faults along which this crustal movement is taking place. The
eroded sediments can be seen pouring from the sides into the valley as
alluvial fans and extending to a great depth beneath the flat floor.

In upland areas, one can often trace the course of tilted strata by
using scarp and dip topography – even where the rocks are mostly
covered by soil and vegetation. This kind of landscape shows which way
the strata are tilted and, when traced across the countryside, picks out
the shape and scale of tectonic folds (see page 41).

Tectonic faults can also be traced in such terrain. Their position
can be deduced where a scarp ridge – that marks the position of a hard
sedimentary rock layer – suddenly comes to an end. This is a clue that
the outcrop of the rock layer has been shifted along a tectonic fault to
somewhere else in the landscape. A geologist investigating this landscape
would then try to locate this particular rock layer on the other side of the
fault line, to work out how much the rocks had been moved along this
fault plane, and in what direction. It is all part of the kind of detective
work needed to work out the history of Earth's crust.

ROCKS AS STORYTELLERS

SURVIVORS:
THE OLDEST ROCKS OF ALL

The first half billion years of Earth is a mystery. No rocks left from that time have ever been found, so there is virtually nothing to go on when attempting to work out the characteristics of the earliest version of Earth. This period is called the Hadean, and in many ways the world would indeed have been a hellish place, certainly for delicate oxygen-breathing animals like humans. For clues as to what there was before Earth formed, we must turn to visitors from outer space – meteorites – for they are debris left over from the building of the planets. We will explore what they represent later (see pages 184–185), but for now we can simply note their timekeeping properties: the oldest of them date back to 4.567 billion years, which approximates to the time Earth began forming.

◀ **The world's oldest rock**
Canada's Acasta Gneiss, at a little over four billion years old, is the oldest rock so far discovered on Earth. During the billions of years of its existence, it has been terribly altered by heat and pressure deep underground, and so gives few clues to the conditions where it originally formed.

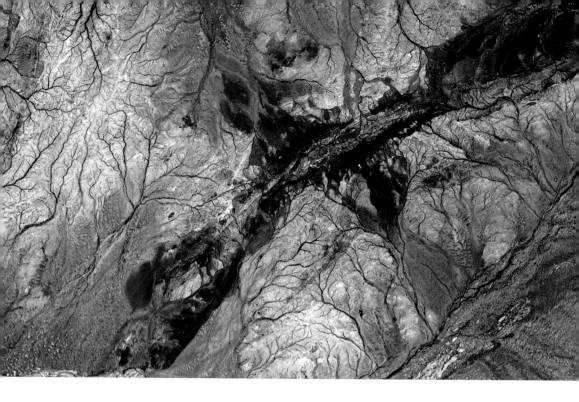

While there are no rocks from very early Earth, a few minerals have been found from that time. They consist of tiny zircon crystals that have been dated to more than four billion years old (the oldest is 4.4 million years). These are all that survive of ancient rocks that eroded, about three billion years ago, into sandy sediment that has since become hard sandstone outcrops in Australia's Jack Hills. What do they tell us, these minuscule Hadean survivors? Chemical clues in them hint that Earth then had some form of crust, with water at the surface. That gives at least some idea of the planet in its infancy.

What about the oldest rock known today? Currently, this is the Acasta Gneiss, originating in Canada. Thought to have first formed as a granite just over four billion years ago, it has been fearfully mangled by mighty tectonic forces in the intervening time, so little of its original character is left. Nonetheless, it is witness to primeval Earth.

Earth's record of rocks mostly began once the mysterious Hadean time had come to an end and the Archean Eon had begun. In the Isua Greenstone Belt of Greenland, there exist sedimentary rocks 3.8 billion years old. These, too, have been altered by heat and pressure – but not so much that they cannot tell us something of that primeval landscape. The climate was warm and perhaps hot. There was no oxygen in the air, so nothing rusted on that landscape. Rivers flowed, and there were seas. And perhaps – the evidence is still being argued over – there was life, in the form of microbes. Very shortly afterwards, convincing clues to life appeared, along with evidence as to how that life was to shape Earth.

▲ **Primeval landscape**
The rocks of the Jack Hills, Australia, seen here in this satellite image, are some three billion years old. Some of them include sand grains of the mineral zircon, which were eroded from yet older terrain. Ranging up to 4.4 billion years old, these are the oldest Earth-formed minerals yet found.

HOTTER EARTH:
ROCKS FROM THE ARCHEAN EON

The youngest Earth of which we have some real evidence –
that of early Archean times from four to three billion years
ago – was very different to the planet we live on today. In
its deep interior, it was several hundred degrees hotter than
today, because it had more inner radioactivity and it retained
more of the heat generated by its rapid, violent assembly from
colliding asteroids and planetesimals.

▼ Relics of a hotter planet
These rock samples
from South Africa,
nearly 3.5 billion years
old, are of komatiite,
a lava that erupted at
higher temperatures
than today's lavas. The
sample on the left has
been weathered by our
current oxygen-rich
atmosphere, while the
sample on the right
shows the large, rapidly
grown 'spinifex' crystals
that are characteristic of
this kind of lava.

One effect of this great internal heat was the eruption of different
kinds of lavas from that of today's. A distinctive one is a komatiite
(named after the Komati River in South Africa), which is common
among Archean rocks but only rarely formed after that. Its unusually
magnesium-rich chemistry means that it erupted at very hot
temperatures – around 1,600°C (2,912°F), compared with normal
basalt lava at 1,200°C (2,192°F). This high-temperature lava was very
runny, flowing almost like water in thin, fast-moving incandescent
sheets across the ground. As this runny lava cooled, large plate-like
crystals grew very quickly, to give the rock a spectacular appearance.

An Earth so much hotter on the inside than today must have
worked differently as a planet. It is thought that the crust of this early
Earth may not have been divided into a number of separate plates,
each moving in a different direction as they do in plate tectonics today.

A DIFFERENT KIND OF PLANET

The very early, hotter Earth had quite a different structure to Earth as we know it today, possibly with a 'one-piece' crust and no plate tectonics.

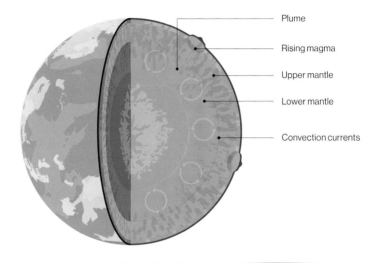

Plume

Rising magma

Upper mantle

Lower mantle

Convection currents

Nevertheless, this young Earth must have found another way to release its internal heat – perhaps by venting magma through 'heat pipes' that punched upwards through a one-piece crust. If there was no plate tectonics, or a different version, the early continents could not have formed (see pages 108–109) above tectonic plates sliding down into the mantle. The Archean rocks preserved include granite-like rocks, through which run basalt-like 'greenstone belts'. These granitic bodies, perhaps precursors to modern continents, might simply have been sweated out of basaltic magmas, while the greenstones might be some kind of early form of ocean crust.

This hotter Earth might have had some surprising consequences, too. At present, there is at least an ocean's worth of water dissolved in the rocks of the mantle. But a hotter mantle can hold less water than a cooler one, the only way to release excess water being to expel it to the surface. If so, the Archean oceans would have contained more water than do contemporary oceans, so Earth might have been more like a water world: mostly submerged, with only the highest parts of the landscape emerging as dry land. This is yet another hint that Earth has not been one constant planet throughout time, but rather a succession of very different planets.

MICROBIAL CONSTRUCTION:
STROMATOLITES

Microbes are tiny and soft, so they do not fossilise easily.
But they can leave traces of their existence in rocks. Geologists
used to wonder at finely layered mound-like structures that
often turned up in rocks so old that they contained no fossils
of familiar animals or plants. It was eventually realised that
these strange layered structures were stromatolites, built by
sticky microbial mats that trapped sediment, layer by layer,
to eventually build up such structures.

The oldest convincing microbial stromatolites are 3.4 billion years old,
found in rocks of the Strelley Pools near Perth, Australia. They can have
a variety of shapes – some look a bit like egg boxes, others like mounds
or columns. They all may represent different forms of microbial colony,
the only kind of life that existed for most of Earth's history.

Today, stromatolites are rare, because on normal sea floors such
delicate sedimentary layers would soon be destroyed by burrowing
or grazing animals, such as snails and sea urchins. However, fine
modern examples can be found in Australia's Shark Bay, on a dry, hot
part of the coast where the seawater is too salty for such animals to
survive. Here, the microbial mats can grow undisturbed and build up
impressive column-like stromatolites. Witnessing them is like taking a
time machine to an Archean seashore.

◀ **Ancient stromatolite**
The typical structure
of curved layers of
a stromatolite. But
beware! – not all curved
sedimentary layers
are stromatolites,
and it takes skill and
experience to recognise
a rock of microbial origin.

Recognising true microbial stromatolites is not always easy, as there are purely chemical or sedimentary processes that can produce similar-looking layered structures. However, in those 3.4-billion-year-old Strelley Pool rocks, not far from where stromatolites were discovered, microscopic fossilised microbes have also been found. The delicate microbial cells were entombed on the Archean sea floor in a fine, soft silica ooze. The ooze layer later hardened into a tough chert rock, which still contains within it outlines of the microbes: some individual spheres, others forming long chains or filaments.

Even where they do not build up to form stromatolites, the mats are sticky and bind grains together, even with the kind of sediment that is normally loose, like sand. When water flows over such a surface of microbial-bound sand, it cannot easily cause individual grains to be swept up in the current; but it can tug at the sticky sand layers, so that they wrinkle up. These wrinkle structures can be preserved in rock strata, to preserve the effects of the sticky microbial mats, even when the microbes themselves have long decayed.

For something like three billion years – most of Earth's history – such structures were the only physical remains within rocks of an abundant, if microbial, life. The kinds of fossils that we are familiar with – shells and bones – were to come much later (see pages 132–133). Nevertheless, this 'simple' early life was to have a giant impact on Earth and its rocks.

▲ **Modern stromatolites**
In the waters of Shark's Bay, Australia, which are too salty to be colonised by the usual seashore animals, microbial colonies build spectacular stromatolite structures, recreating a Precambrian seascape.

CHANGING ATMOSPHERICS:
THE OXYGEN TRANSFORMATION

The role of oxygen on Earth is strange and seemingly contradictory. Oxygen is the most common type of atom in Earth's crust, making up most of the common rock-forming minerals (quartz is SiO_2, for example). However, for the first two billion years of Earth's history, there was virtually no oxygen in the atmosphere, because the pure gas is so chemically reactive, quickly forming new chemical compounds.

On an early Earth without free oxygen, the oceans soon filled up with dissolved iron, and became like the toxic, iron-rich waters found today in old underground mine workings. From about 3.7 billion years ago, this iron began to precipitate out onto the sea floor to form huge deposits of iron-rich strata: the banded iron formations. These are striking to look at, with thin layers of bright-red iron oxide separated by white silica-rich layers. They are by now the most important iron ores on Earth – almost all the iron used will have come from them. The strata probably formed with the help of microbes, perhaps using an early kind of photosynthesis that did not release free oxygen into the atmosphere, but bound it to the iron in the ocean instead.

▼ **Iron outcrop**

Thick deposits of banded iron formation strata at the Hamersley Gorge in Australia, show the scale of iron removal from the oceans 2.6 billion years ago.

About 2.5 billion years ago came the first real signs that oxygen was building up in the atmosphere. This is indicated by a colour change to the sediments that formed on Earth's land surface and the type of minerals they contained. In times before oxygen came into the air, that sand could contain grains of minerals such as pyrite (fool's gold). However, when oxygen gas appeared, it chemically attacked the pyrite and other similarly reactive minerals, and turned them into iron oxides and hydroxides – that is, to rust. That rusting would have turned the whole landscape from shades of grey and green into a palette of reds, browns and oranges. This transformation can now be seen as the appearance of the strata known as 'red beds', which contain an abundance of such rusted minerals. From that time, 'red beds' have been an important part of Earth's sedimentary record.

The appearance of free oxygen in the atmosphere and oceans was to cause enormous changes to Earth. It led to an 'oxygen catastrophe', as most microbes then would not have been able to tolerate this fiercely reactive gas – as toxic to them as chlorine gas is to humans today. These ancient microbes retreated to oxygen-free refuges, such as occur in stagnant swamps or deep underground. Other microbes, though, evolved to cope with this new, more energetic kind of chemistry: eventually, this led to the kind of complex oxygen-respiring life seen today, and this would go on to change rocks in its own fashion.

▲ **Oxygen signal in strata**
The red bands in the John Day National Monument Park in Oregon are fossilised soils, some 30 million years old, and testimony to oxygen in the atmosphere that turned surface iron minerals into rust. Strata like these first widely appeared on Earth a little under 2.5 billion years ago, when the atmosphere became oxygenated.

FORCE OF NATURE: HOW ANIMALS CHANGED EARTH'S GEOLOGY

▼ Early animals make their mark

The trilobites (left) and brachiopods (right) were among animals that appeared as part of the 'Cambrian Explosion' of life. Their fossilised remains (and, in the case of the trilobites, traces of their walking and digging activities) are a marker of rock strata formed during this time.

It is only in the last one-eighth of the history of Earth that all the main groups of animals now familiar to us – snails, worms, molluscs and so on – appeared. They quickly dominated life in the oceans and, somewhat later, on land, and left abundant fossil remains in rock strata. To early geologists, this sudden appearance of abundantly fossil-rich rocks, after the great stretches of older rocks with few or no fossils, was so striking they called it 'the Cambrian Explosion' (used to mark the beginning of the Cambrian Period of geological time). With closer study, it became clear that this appearance of complex multicellular life forms was not actually abrupt, in reality being a burst of evolutionary change starting about 550 million years ago and lasting some 30 million years. Nevertheless, it was another great transformation of our planet – and it transformed the nature of its rocks, too.

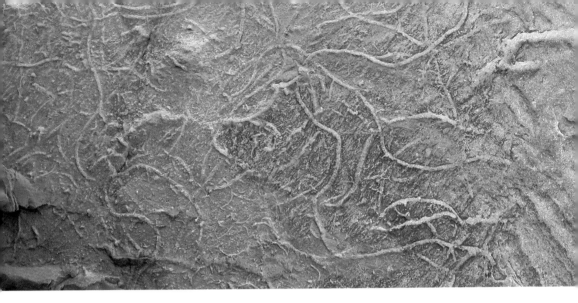

One change was in the nature of sedimentary strata. Many of these newly evolved animals were muscular and highly mobile, often moving to either hunt other animals or to escape from animals that were hunting them. Many, as they moved, either walked or slithered across the sea floor, or wriggled down into it. As a result, they punctured and churned sea floor sediments that previously had been undisturbed, or only stirred up by waves and currents to produce structures like ripples and dunes (see pages 84–85). This kind of biological disturbance was new, and produced a new set of marks in the strata: footprints, crawling traces, burrows – sometimes so energetic as to completely mix the original sedimentary layers. Geologists call it bioturbation. It is almost completely absent from strata older than 550 million years old and abundant afterwards.

Many of these animals also had skeletons, usually of calcium carbonate or phosphate, for defence, attack, or simply as a framework for muscles. These hard skeletons, once buried in sedimentary layers, could be preserved as obvious, complex fossils – a key marker of rock strata of the last half billion years. If enough skeletons piled up, they could form most or all of rock layers in their own right, as in strata of shelly limestone, and – though somewhat more infrequently – bone beds (see pages 94–95). Fossil plants can have skeletons as well, and can also build up rock layers in their own right (see pages 142–143). But when it comes to piled-up animal skeletons, the largest and most spectacular examples are provided by reefs.

These new kinds of sedimentary rock are such a distinctive feature that this time interval has been given the status of an eon, the largest kind of geological time unit: the Phanerozoic Eon, of which the Cambrian was the first geological period (the Cambrian has long finished, but we still live in Phanerozoic time).

▲ **Petrified pathways**
Fossilised trails made by worms and other animals crawling on an ancient sea floor, now preserved as a bed of sandstone.

SITES OF DIVERSITY:
ANCIENT CORAL REEFS

To a mariner, a reef is a dangerous set of rocks near a coastline, where shipwreck threatens. To a biologist, a reef presents a marvellously diverse ecosystem, where corals provide structure and shelter to a dazzling array of other living organisms. However, to a geologist, a reef means rock units that have been produced by its living organisms, sometimes on a gigantic scale over geological eras.

Charles Darwin saw this during his round-the-world voyage on the *Beagle*, and marvelled at how animals as delicate and translucent as corals could build up massive units of limestone rock, as generation after generation of their skeletons piled up. And while the coral skeletons provided the main reef structure as a kind of 'hyper skeleton', the skeletons of many other organisms, such as bivalved molluscs, snails and 'living rocks' of calcareous algae, added to the structure.

Darwin also had the brilliant insight that coral reef rocks can stretch across thousands of kilometres and grow enormously thick, producing giant mountains of reef limestone. He puzzled over strangely ring-shaped coral atolls, and worked out that they had started growing around an island volcano. Then, over geological ages, the volcano slowly sank, while the corals continued to build up in the sunlit shallow water where they flourished. Finally, the volcano was entirely submerged, leaving just the ring of corals at the sea's surface. Years later, Darwin's idea was proved correct, when an atoll was drilled into to reveal more than a kilometre's thickness of coral limestone before the ancient buried volcano was reached.

◀ **Coral rock**
This fossilised coral colony, by the Red Sea, is of the kind that helps build the limestone substructure of a reef.

REEF ANATOMY

Each zone of a reef will have its own kind of biological community, and this will be reflected in the patterns of fossils found in an ancient reef limestone.

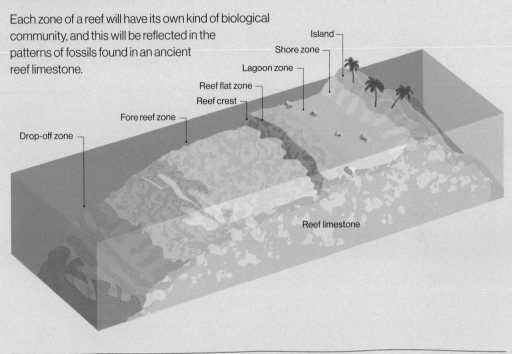

Island

Shore zone

Lagoon zone

Reef flat zone

Reef crest

Fore reef zone

Drop-off zone

Reef limestone

Corals today are the main reef-builders (though it is the more durable coralline algae that form reef fronts, against which the waves crash). In earlier geological times, other animals built up reefs and produced different kinds of reef limestone. In the Cambrian Period, more than 500 million years ago, extinct vase-shaped organisms known as archaeocyathids built small reefs on the sea floor. And some 400 million years ago, in Devonian times, huge reefs existed, where sponges were important reef-builders. Some of the most bizarre reefs grew 100 million years ago in Cretaceous times, when dinosaurs lived. These were formed of bivalved molluscs called rudists, which had evolved to a tubular shape and grew from the sea floor, packing together to form reef structures. All these types built both complex reef ecosystems and reef rocks using different biological components.

Reefs are delicate ecosystems that have waxed and waned throughout geological time, sometimes disappearing completely for millions of years in what is known as 'reef gaps', when environmental conditions become too difficult, often caused by climate change. Today, reefs are under threat once more from human-made global warming.

DESERTSCAPES: DUNES, FULGURITES AND SALT ROCKS

▲ Ancient desert dunes
These Navajo sandstones, found on the Colorado Plateau, are the petrified remains of wind-blown desert dunes that covered part of North America's Midwest in Jurassic times, 190 million years ago. The shape of the strata represent the way in which these ancient sand dunes grew and migrated.

For much of Earth's history, most of the land was a desert, without forests, grasslands or deep soils beneath these ecosystems. Then, after Earth's land surfaces greened (see pages 142–143), the desert regions retreated to places too hostile for life, mainly the harsh, arid regions of the subtropics, a landscape akin to the Sahara Desert today.

The most characteristic desert strata are the petrified remains of huge sand dunes, preserved as wind-driven patterns of cross-bedding (see pages 84–85). Very rarely, a whole dune, many metres high, can be fossilised, for example, when covered by a thick armour of lava in a volcanic eruption. However, usually the tops of the dunes were blown away before being buried under more desert sand, so that their bottom parts are left, the tilt of their layers betraying the direction from which prehistoric winds blew. Spectacular examples exist in the Navajo sandstone of the Colorado Plateau, formed in Jurassic times.

The sand grains themselves have been shaped by their journey across the ancient desert, their surface rounded and frosted by countless impacts with other sand grains; they are beautifully size-sorted, and have been winnowed clean of mud and mica flakes by wind. Now and then, one finds within them little vertical cylinders of fused silica: these are fulgurites, formed by lightning strikes during electric storms.

In these hot, arid regions, lakes and arms of the sea that form during rainy seasons quickly dry up, giving layers of different kinds of salts: gypsum, sodium chloride ('common salt'), sodium sulphate of bitter lakes and – from the most concentrated of brines – potassium and magnesium chlorides. Dried-up lake deposits of deserts often show such salt layers, their bright whiteness contrasting with red lake mudrocks.

On a much larger scale, entire enclosed seas in the world's arid zone can dry out, leaving colossal salt deposits, or 'saline giants'. One classic example occurred in the Miocene Epoch, when the entire Mediterranean Sea dried up. In the strata deep beneath the present-day sea, Messinian salt deposits are 2 kilometres (1.25 miles) thick, resulting from the sea filling and drying out many times over 500,000 years. This single episode locked away enough salt to make the world's oceans 5 per cent less salty! Some of the salt layers, such as spectacular gypsum 'forests', have been lifted to the surface in places like Spain.

The thick salt layers are soft and of low density; when pressed down by trillions of tons of overlying rocks, they can break through their rock roof and slowly flow upwards in monstrous columns, puncturing and pushing apart the overlying strata. These salt domes show the dynamism of the underground rock realm.

▼ Fossil lightning strike
The dark structure running down the middle of the photograph is a fulgurite, made of melted sandgrains, which was produced by a lightning strike on desert sands 300 million years ago.

WATERBORNE: HOW RIVERS DISPERSE SEDIMENT

Rivers have always flowed on Earth, ever since our planet cooled enough for rain to fall onto the land surface more than four billion years ago. However, there is much more to rivers than the flowing of river water in its channels: the sediment the waters carry forms distinctive types of rocks. There is even a type of river that flows beneath the sea to build strata.

▼ River channel maze

Iceland's Jökulgilskvísl River is a classic example of a modern braided river, with many constantly switching channels. Sand and gravelly strata from rivers such as these are common in the geological record.

We can still see the most ancient kinds of rivers, which typically flowed on a primitive Earth before vegetation spread across the land to change the way rivers behaved. These occur in modern desert areas, and in glaciated mountainous and polar regions. Here, the river water splits into many ever-changing shallow channels, separated by shifting banks of sand and gravel. In the 19th-century United States, gold miners heading west picked their way across some of these, cursing them as 'a foot deep and a mile wide'. Called braided rivers, in ancient strata the shapes of the bars can still be seen, the pebbles often neatly stacked against each other by a flowing current.

Today, the most familiar rivers have one main channel – an efficient way for water to flow across most landscapes, with their thick soils held together by myriad plant roots. These meandering rivers have a sinuous channel that, over time, migrates from side to side across a wide floodplain. As it does so, it leaves a layer of sediment, typically with gravel at the bottom (from the river channel floor), to sand with gently inclined layers (from the sandy point bar on the inside of meander bends), to mud at the top (from floodwaters that spill over the floodplain). This pattern of river rock strata formed more commonly on Earth, after the land surface greened about 350 million years ago, but can also be found in unexpected places, such as on Mars (see pages 198–199).

Rivers that flow under the sea are often a submarine extension of today's rivers, but with a different flow pattern. Offshore from the river mouth, masses of river sediment can slowly build up on the sea floor. These sediments can then be destabilised suddenly by a storm or earthquake that triggers a turbidity current, which then flows downslope under the power of gravity, eventually dumping a layer of turbidite sediment on the deep-sea floor (see pages 90–91). During its journey, though, the turbidity current can flow within a submarine canyon, or form a meandering channel along the deep-sea floor. The strata formed by these extraordinary submarine rivers can be detected by sonar on modern sea floors, while ancient examples can be found among the rock strata of mountain belts.

▲ **The meandering Mississippi**

A classic example of a meandering river, the Mississippi's channel changes its position over time. The yellow line is the Arkansas-Mississippi boundary, drawn along the river when its position was different.

ANCIENT COASTLINES:
HOW THEY'VE CHANGED

Today's pattern of land and sea seems changeless and eternal; the shapes of islands and continents have remained much the same for many human generations. But if we travel further back in time, there is evidence of great change. Twenty-thousand years ago, when the last ice age was at its peak, the world's ice caps had grown so large, and taken so much water out of the oceans, that the sea level was 130 metres (427 feet) lower than today. Humans could walk across dry land from Siberia to Alaska and the Americas, and in Europe, they inhabited a landscape now under North Sea waters.

Even further back in time, coastlines changed continually: sea levels rose and fell, the crust of the land was uplifted or subsided, and the continents themselves, caught up in plate tectonics, changed position on the globe. One can see abundant evidence of such change in the rocks around us: much of what is now dry land consists of strata that formed on a sea floor containing countless fossils of marine animals.

One of the most profound expressions of such change is where a mountain chain is worn flat by erosion, and much younger strata are deposited on top of the eroded surface. The plane along which they meet is called an unconformity, and represents a time gap of perhaps

▼ **Coastal change**
When the sea level falls (or, more commonly, when part of Earth's crust rises), a beach and the cliff behind can be lifted beyond the reach of the sea, and a new cliff and beach will begin to form in front of it.

hundreds of millions of years. Recognising an unconformity in Scotland was the key that allowed 18th-century scholar James Hutton to first sense the enormity of deep geological time. Hutton realised that this kind of contact between rocks meant that there must have been sufficient time for a mountain belt to grow, slowly erode to its roots, and then be buried under newer strata – which in turn were uplifted to form a new landscape.

Many unconformities have now been recognised in Earth's geological record. A notable one is the 'Great Unconformity', which formed half a billion years ago, when sandstone strata shaped on a shallow sea floor overlaid the deeply eroded rocks of much older landscapes. The erosion that led to it may have been caused by a worldwide glaciation.

Not all transitions between land and sea are so abrupt. If sea level rises progressively across soft low-lying landscapes, marine strata can overlay sedimentary layers formed immediately before on land. A beautiful example can be seen in southern Great Britain, marking where the Jurassic sea swept across a landscape of rivers and coastal plains formed in the preceding Triassic. The change from red mudstones made on land, to pale lagoon muds, to dark mudstones full of marine fossils such as ammonites, is striking.

▲ **Hutton's unconformity**
Siccar Point, Scotland, where James Hutton recognised gently inclined red sandstones resting upon an eroded surface of much older, near-vertical strata, and realised that an immense span of time must separate these two rock groups.

FLORAL EXPLOSION:
THE GREENING OF THE LAND

Life began on land three billion years after microbial life emerged, more than 100 million years after the Cambrian Explosion (see pages 132–133), when a dazzling variety of large, complex animals evolved to colonise the oceans. However, compared to the constant and supportive environment of the sea, the land is a difficult place for life, with huge swings in temperature and the danger of drying out. Adapting to its challenges took a long time, as the land itself needed reshaping to make it more habitable. Here, plants were key.

The first green shoots can be seen in river-laid rock strata (see pages 138–139) from 430 million years ago, their fossilised remains seen now as thin carbonised lines of matchstick width and a few centimetres long. They are just shoots: slender branching stems, with little spore

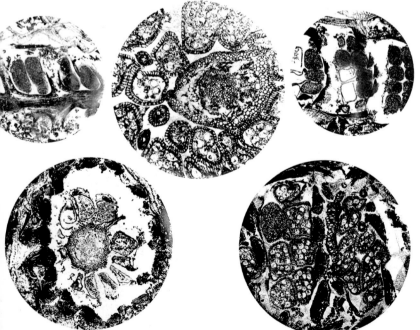

◀ **Perfect plants**
Coal seams can include beautifully preserved plants, which were naturally impregnated with calcium carbonate just after death. These microscop images show the kind of cellular detail of that can be seen in such remarkable fossil material.

cases, from a plant called *Cooksonia* that could have lived only in damp, sheltered patches near lakes and rivers.

Over the next 100 million years, through the Devonian Period, into the Carboniferous, these land-formed strata showed an increasing complexity of plants as they evolved better means to adapt to land and to spread more widely across it. Roots appeared, as well as branches and leaves, seeds, pipe-like tissues to conduct water, and tough outer tissues to prevent water loss.

By the middle of the Carboniferous, plants had grown enormous, with giant horsetails, ferns and conifers forming thick forests. An accident of geology led to the preservation of many such forests, when thick seams of a new kind of rock consisting of fossil plants appeared: coal. The formation of coal pulled so much carbon out of the air and buried it underground that global climate cooled, and a 50-million-year ice age began, with giant ice caps forming on South America and southern Africa (then, the South Pole). Now that humans burn this coal and return the carbon to the air, the process is reversing and heating up Earth's climate.

The new plant life was an enormous expansion of the biosphere, persisting today (forests represent the bulk of life on Earth by weight). The plants did not just live on the landscape, they re-engineered it. Their roots bound the land's surface sediment together, while decaying remains fertilised it to make thick soils. In so doing, they changed the ways that rivers flowed (see pages 138–139).

The spread of plants on land provided food, shelter and expansive living space – animals could move onto and colonise land. This animal invasion followed that of plants so closely, that its record in rock strata looks almost simultaneous, with tiny invertebrates and bizarre armoured fish spreading widely into Devonian lakes and rivers. When Carboniferous-era coal swamps evolved, amphibians and the first reptiles began hauling themselves up onto land, along with dragonflies with 60-centimetre (24-inch) wingspans and millipedes 2 metres (6.5 feet) long. The rocks show breathtaking evidence of Earth's transformation.

▲ **Pioneer lungfish**
This lungfish was part of the invasion of the land by plants and animals. It lived in a large lake in what is now Caithness, Scotland, 390 million years ago. The lake's surface waters were oxygenated, allowing these and other early fish to thrive, while the lake floor was stagnant, which helped fossilise their remains after death.

MARINE CATASTROPHE:
WHEN OCEANS DIE

Oceans can die in different ways. Animals and plants that live in them can simply become extinct, most often due to being starved of oxygen. This usually happens when Earth's climate heats up too much, so that ocean circulation slows or stops, with oxygen no longer carried into its depths. There have been many examples of such 'ocean anoxic events' in Earth's history, some worldwide and reaching even into shallow water. In the greatest of these, mass extinction events have taken place, including the Great Dying, when, 250 million years ago at the end of the Permian Period, some 95 per cent of species of life on Earth were killed off. The rocks produced during these episodes are characteristically dark, full of carbon from decaying organic matter, and can contrast strikingly with rocks around them. Ironically, these dark organic-rich mudstones have yielded much oil and gas: but by burning these, we are generating another global warming and ocean oxygenation crisis.

▲ **Oxygen crisis strata**
This bright white rock is chalk dating back to the Cretaceous Period, which formed about a hundred million years ago as oozes on a sea floor rich in animal life. The dark layer represents a time of many thousands of years, when the sea floor became starved of oxygen and this animal life died out.

Oceans – or an isolated part of them – can die by losing their water to evaporation as well, as happened to the Mediterranean Sea between five and six million years ago (see page 137). But here, the water can return, just as oxygen can return to an anoxic ocean when climate cools and ocean circulation begins once more.

But the most permanent way of killing an ocean is by physically removing the sea floor and the ocean crust rocks beneath it. Thanks to the action of plate tectonics (see pages 28–29), this is happening – steadily and inexorably – all the time.

The Mediterranean Sea is a classic dying ocean. This is the shrunken remains of a once-mighty ocean a few thousand kilometres wide called the Tethys Ocean. The Tethys was caught, as in a gigantic vice, between the colliding continents of Africa and Eurasia, and its ocean crust was pushed into Earth's mantle. One result of this collision is the Alpine chain of mountains. The ongoing continental collision, though, is untidy. Not all of the Tethys ocean crust has been subducted into the mantle, but some fragments were obducted (scraped off) and pushed up onto the land. These detached pieces of ocean crust rock are called ophiolites; fine examples exist in Oman and Cyprus, made up of ocean floor basalts, shot through with igneous dykes, and now high and dry on land so they can be easily studied.

There are many places on land that mark vanished oceans. The remains of the Iapetus Ocean, which existed 500 million years ago, can be traced from Europe into North America. In Great Britain, it ran between modern England and Scotland, and on through Newfoundland into what are now the Appalachian Mountains. In these places, scraped-off and slid-together slabs of deep-ocean strata, with fossils, can be traced along the hillsides, to help picture this ancient ocean.

▼ **Bright stripes reveal oxygen at ancient sea floor**

The thin pale striped areas are the tops of mud layers that, 420 million years ago, during the Silurian Period, were found on the sea floor, where they lost part of their carbon content in a reaction with the oxygenated seawater immediately above. They give insight into the sea floor of that period – close examination shows small animal burrows within it as additional evidence.

DEEP IMPACT:
WHEN ASTEROIDS STRIKE

Compared with earlier times, few asteroids now strike Earth. Most of the traces of an early ferocious bombardment have been lost, churned over by tectonics or eroded away. But its scale can be seen by looking at the battered face of the Moon, which still bears the scars of these huge early impacts. Damage to Earth must have been greater: its larger mass would have attracted many more meteorites.

When meteorites do strike Earth, few are large enough to do damage. However, 50,000 years ago, a meteorite 30 metres (99 feet) across struck the Arizona desert at a speed of 16 kilometres (10 miles) a second, devastating the local region and leaving a crater 1.2 kilometres (0.75 miles) across, leaving masses of rubble inside and outside. This rubble is an impact breccia, where pulverised and melted fragments of blasted-out rock are mixed with droplets of melted iron from the meteor.

Thirteen million years ago, in the Miocene Epoch, a larger meteorite struck what is now north Germany, affecting a local but not global area. It left a crater some 24 kilometres (15 miles) across: the old walled town of Nördlingen was later built in its centre. This colossal impact produced a rock made of yet more finely pulverised and melted impact debris called suevite, which has been used as a building stone locally.

▼ **Meteor Crater, Arizona**

Most of this 50,000-year-old iron meteorite vaporised on impact, though fragments can be found among the debris. The dry climate has helped preserve the crater, though its rim has been lowered by erosion, and the crater floor is being slowly filled by post-impact sediments.

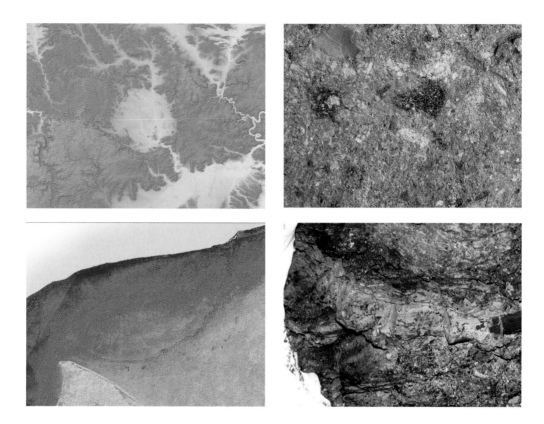

Sixty-six million years ago, a large impact did change the course of Earth's history, setting the pattern for today's world. The first evidence to emerge was a profound mass extinction in which dinosaurs and other life forms disappeared. Then, a thin rock layer rich in iridium – more common in meteorites than on Earth – was found throughout the world, which had formed alongside the mass extinction. This led scientists to suspect that a giant meteorite impact had caused the catastrophe – but where was the crater? A huge crater, 200 kilometres (124 miles) across, of exactly this age was later found, buried deep in the ground below Mexico, made by a meteorite an estimated 10 kilometres (6 miles) across. All the evidence fell into place, and the impact is now widely accepted. In the crater itself (which formed into a shallow sea), there is a suevite layer 100 metres (328 feet) thick. Along neighbouring coasts, there is a sedimentary layer swept by the mighty tsunami unleashed by the impact. And farther away, extending worldwide, there is that thin layer rich in iridium, and also in tiny melted droplets and fragments of impact-shocked mineral. This rock layer encircles the world – and also signals its dramatic change.

▲ Impacts and products

The meteorite impact crater at Nördlingen, Germany, still forms a visible, rampart-rimmed depression in the landscape (top left), filled with the blast-fragmented rock suevite (top right). The Chicxulub crater in Mexico, responsible for the demise of the dinosaurs, is buried beneath later strata but may be detected by geophysics (bottom left), generating an iridium-rich layer that was deposited worldwide (bottom right).

HOTHOUSE: THE ROCKS OF A WARMER EARTH

The current ice age exists as what would normally be a brief warm interval before the ice returns. In today's reality, though, humans are warming Earth so quickly through burning fossil fuels that the ice may not return for many millennia, and the climate may be pushed into the 'hothouse' world dinosaurs lived in, when carbon dioxide concentrations were twice or more that of today's. So, what kind of rocks mark a much warmer Earth, as a future guide?

The world then did not have thick ice caps covering Antarctica and Greenland, and the sea level was about 100 metres (328 feet) higher than today. Therefore, much of the world's continental area was flooded by sea, its floor buried by thick layers of white ooze consisting of microscopic skeletons of planktonic algae. This is the chalk layer, a key rock marker of a warmer Earth.

Look closely at a chalk cliff to view subtle patterns of striping, with layers of whiter and greyer chalk. In them can be seen the workings of subtle, regular climate changes on that hothouse Earth, as weather and ocean current changes affected the microscopic plankton of the chalk

The layers visible in this cliff represent repetitively changing climate within a global greenhouse climate, driven by rhythmic changes in Earth's orbit and spin. Each layer represents a climate cycle some 20,000 years long.

▼ **Lush polar life**

In times of a greenhouse world, regions of Earth that are now frozen and barren were temperate, with lush forests and many animals, including dinosaurs.

ocean, so changing the composition of rock layers. Those changes in turn were controlled by regular alternations in sunlight patterns falling on Earth, as Earth's spin and solar orbit slowly changed in cycles lasting tens of thousands of years. That astronomical pattern provides a chronometer for the rocks, to calculate the many millennia that chalk layers needed to form.

The nature of a hothouse Earth is best seen in the polar regions, wherein lie the most dramatic differences from our current world. Antarctica in the Cretaceous lay over the South Pole as it does today, yet the fossil-rich Cretaceous rock strata there show that it was covered not by ice, but by thick rainforest of tree ferns and conifers. These were strange rainforests, for there would have been continuous daylight in summer and darkness in winter, as with the polar regions today. Nevertheless, this lush vegetation flourished, so much so that coal layers formed from their buried and compressed remains.

This provides a vision of what Antarctica might return to, if fossil fuels continue to be burned – though the price for the greening of Antarctica would be the drowning of landscapes elsewhere from ice melt.

What about the tropics? Here, temperatures did not rise quite so much in past greenhouse worlds, but the ocean surface could become too hot for plankton to thrive. With fewer plankton skeletons, less limestone rock was formed. Then, when carbon dioxide levels soared, the oceans became more acid, so many plankton skeletons dissolved before they could form rock layers. Today, we are seeing a return to this rock-dissolving world.

ICE AGE: ROCKS IN A COLDER ERA

Earth has gone from eras of global hot climate, with little or no ice, to times of bitterly cold climate. Ice covered the entire planet 700 million years ago, turning it into a 'snowball' Earth: from outer space, it would have resembled a frozen moon of Jupiter or Saturn.

Currently, we live in a brief warm episode within a less-extreme ice age, when the ice has retreated to cover only Antarctica, Greenland and high mountaintops. But 20,000 years ago, ice spread much more widely, to reach places as far south as New York.

The passage of billions of tons of ice leaves many marks on rock surfaces. These ice scratches can be found on today's mountainsides, left by glaciers thousands of years ago. They can be found in ancient rocks as well, left behind by the ice ages of millions of years past.

Ice can make rocks, too. As it moves, it scrapes together and mixes large masses of sediment and rock fragments, spreading these as thick layers of boulder clay ('glacial till') across the landscape. Finding this kind of deposit is an unmistakable sign of past glaciation. Geologists

▼ **Dynamic rocky landscape**
The rocks of this landscape in Patagonia, Argentina, are being scraped and eroded by a glacier. The small icebergs breaking off are carrying the resulting sediment down to the sea where, drifting to the ocean floor, it will form the beginnings of new rock strata.

seek ancient, hardened examples of glacial till (tillite) to reconstruct the glaciations of Earth's deep geological past, including of the ancient 'snowball Earth' glaciations originally revealed by the discovery of thick tillite strata on continents then located at the Equator.

As glaciers and ice sheets melt, water streams out of them. This can take place steadily, to form seasonal glacial rivers (flowing in summer, freezing up in winter) that winnow the boulder clay to leave large deposits of sand and gravel. These deposits are now avidly sought as construction materials (see pages 162–163). Glacial meltwater can also produce catastrophic floods, especially when it pounds behind giant ice dams, which then break to produce gigantic torrents that scour the land downstream.

Reading the history of the ice ages from strata such as glacial till, sand and gravel is not easy, as they are so patchy and as easily removed by erosion as they are to deposit. A more complete indicator of ice age history can be found beneath the deep ocean floor, in the oozes slowly and steadily accumulated there: these contain chemical clues to the twists and turns of ice age climates. Deep-sea drilling has revealed that there were not just four major glaciations in the last 2.6 million years, but more than 50, separated by warmer interludes. These climate fluctuations, like those seen in the Cretaceous chalk (see pages 148–149), were paced by changing patterns of sunlight on Earth, as Earth's spin and orbit slowly and rhythmically changed.

▲ **Ice-scoured rock**
This rock surface has been scratched and grooved as an advancing glacier has slid across it. The finding of such 'glaciated pavements' preserved in regions that are ice-free today was a key piece of evidence, showing that ice ages had occurred in the past.

POLAR RECORD: WHAT ICE CORES TELL US ABOUT CLIMATE

Ice is actually a rock, albeit one that melts at temperatures comfortable to humans. It is a major rock on moons such as Europa and Titan in the frigid outer solar system (see pages 202–205). On Earth, ice strata only occur in the polar regions and on high mountaintops, where they tell eloquent tales of an ancient Earth.

Earth's main ice strata cover Greenland and Antarctica, where they can approach 5 kilometres (3 miles thick) in some places. This ice mass starts off as fluffy snow falling on the ice-sheet surface. In many places, it is so cold that it does not melt, even in summer, and the snow layers pile up year after year, in time compacting into hard, solid ice.

Eventually, the ice becomes so thick, it begins to slowly flow under its own weight towards the edge of the ice sheet, where it breaks off as icebergs and floats off to melt back into the ocean. But by this time, an enormous, continuous archive of ice strata can pile up. Drilling into

▲ **Annual snow layers in Antarctic ice**

The snow captures clues from the surrounding environment, such as dust and trapped air, which can be used to reconstruct climatic and other history.

the centre of the Antarctic ice sheet has revealed ice layers there that back 800,000 years (the Greenland Ice Sheet dates back 120,000 years). The ice cores extracted in this drilling reveal extraordinary stories of Earth's past climate, including unique information such as a record of fossil air (not all of the air is squeezed out as the snow compacts down; myriad air bubbles are left in the ice).

Analysing this fossil air has shown that previous levels of carbon dioxide have varied rhythmically from about 180 parts per million when the climate was very cold (the temperature is calculated from the chemistry of the frozen water), to about 280 parts per million in warmer interludes. These ice strata provide the crucial 'baseline' record of carbon dioxide in the atmosphere, illustrating the scale of human change to this greenhouse gas through burning fossil fuels (now at over 400 parts per million and rising).

The ice strata tell other stories as well. The snow surface acts as a kind of flypaper to catch other particles blown through the air. It catches dust, showing that the air in glacial times was much drier and dustier than in warmer interludes. It catches far-drifted volcanic ash and sulphur, indicating when major volcanic eruptions took place. And more recently, it has trapped minute amounts of lead (from lead smelting dating back to pre-Roman times, and more recently from lead formerly used in petrol as a car fuel). It is an amazing historical tapestry that will last for as long as these ice rocks survive.

▼ **Climate history from ice layers**

Systematic analysis of fossil air from Antarctic ice shows rises and falls between low and high levels of carbon dioxide that coincide with glacial and warm intervals of the ice ages over the past 800,000 years.

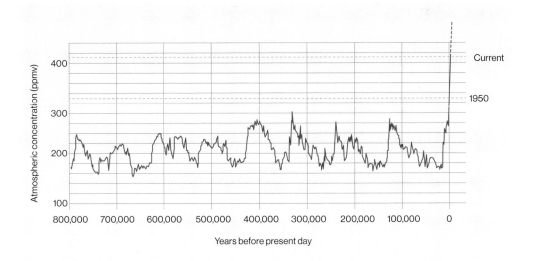

HUMAN-MADE ROCKS

EXTRACTION:
QUARRIES AND MINES

The Stone Age may have been a long time ago, but we now use more rocks than ever. To satisfy our daily needs, everything we do not grow must come out of the ground. Bricks, concrete, steel, copper, oil and coal, plastics, materials that computers and smartphones are made of – all must be taken from rocks underground. The most direct way to extract them is to dig them out, either by making large holes in the ground's surface, in quarries and open-pit mines, or by burrowing to form underground mines.

The scale of this enterprise is now gigantic. If we take just the basic material, rock itself, either as crushed rock or loose sand and gravel for roads, foundations and building material, then collectively we extract 50 billion tons a year. This is roughly 7 tons for every person on the planet, or enough to build 8,000 Great Pyramids of Khufu each year!

This kind of rock extraction is carried out in half a million pits and quarries worldwide, for rock, sand and gravel are almost omnipresent: virtually all material extracted can be used. Most of the very large, deep holes in the ground are dug to extract more uncommon materials, where a large mass of rock yields a small amount of resource. For example, most copper ores mined contain less than 1 per cent of copper – the rest is waste rock that is thrown away (for an average house with 91 kilograms (200 pounds) of copper wiring, 10 tons of rock waste is generated). Copper mines can be gigantic –

the Bingham Canyon open-pit mine in Utah is 4 kilometres (2.5 miles) across and 1.2 kilometres (0.75 miles) deep. For diamonds, the amount of waste is even greater: 7 tons of waste rock is typically extracted to obtain two carats of rough diamond, which ends up as one carat (one-fifth of a gram) of polished diamond. As diamond ore is found in volcanic pipes, the resulting mines are often spectacular (see page 67).

Underground mining is hidden from us, but its scale is equally breathtaking. Such mines have been excavated by humans since the Stone Age (to extract flint for tools), and have expanded enormously since: for coal, metals and other resources. A key step in the Industrial Revolution was the invention of steam engines, to pump water out of mines (that would otherwise rapidly fill with water), allowing them to go much deeper than before. Now, the deepest mines in the world (gold mines in South Africa) can be 4 kilometres (2.5 miles) below ground. These must be cooled artificially, and engineered to prevent 'rock bursts' at such immense pressures.

Although the mines themselves are invisible to us, their effects can be seen, as these huge underground voids subsequently collapse and form subsided areas in the landscape.

DEEP MINING AND SURFACE EFFECTS

The extraction of coal seams deep underground affects surface terrain in that, when the mine workings eventually collapse, areas of subsidence occur. Earthquakes (usually small) can be generated by underground collapse events, while groundwater circulating through the old workings can become iron-rich and polluted.

Within the landscape, broad troughs or more restricted sinkholes can form when one or more of the pillars holding up the mine roof give way, with the rock above collapsing down into the void. More recently, mines have had such collapse built into the design, to allow better control of this inevitable subsidence.

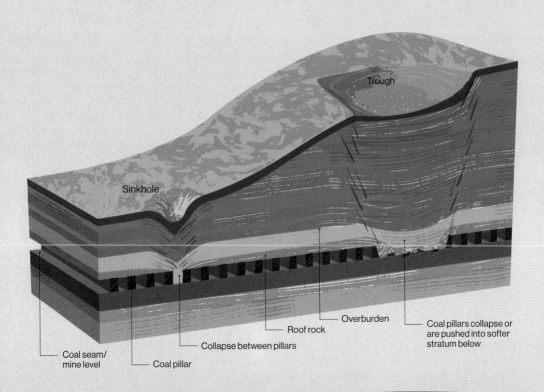

Trough

Sinkhole

Overburden

Roof rock

Coal pillars collapse or are pushed into softer stratum below

Collapse between pillars

Coal seam/ mine level

Coal pillar

MINED AND MANUFACTURED:
NATURAL AND SYNTHETIC MINERALS

Earth has long been a mineral paradise, probably more so than any other planet or moon in the solar system. This wealth comes from its complexity and variety as a living planet. Outer space is far more impoverished: astronomers have identified just a dozen or so minerals in the cosmic dust expelled from dying stars, which comprises the starting point for all minerals and rocks in the universe. Meteorites – fragments of the building blocks from which the solar system formed – have yielded just some 250 minerals.

Once a planet begins formation, though, processes such as magmatism and metamorphism cause new chemical combinations to materialise, creating a range of new minerals. On an early Earth, before life arose, there may have been around 2,000 minerals. Some geological environments make more minerals than others. A particular type of natural Aladdin's cave is found in pegmatites: mineral veins in granites that concentrate rare elements. Here alone 500 minerals have been identified. Later, when life emerged, more minerals formed. A key step took place about 2.5 billion years ago, when photosynthetic plants evolved, and the atmosphere become oxygenated, which led to the formation of many oxides and hydroxides. By that time, Earth contained some 5,000 minerals – an inventory that stayed more or less stable until recently, when humans began to alter the constituents.

▼ **Natural beauty**
Quartz is a common rock-building mineral on Earth. Shown here is an uncommon variety, ametrin, which combines the characteristics of amethyst and citrine.

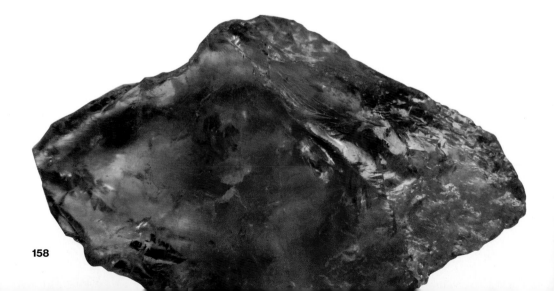

One of the first changes took place a few thousand years ago, when humans began to separate metals, such as copper, tin and iron, from ores. The pure metals are rare in nature, and they were extracted in larger amounts to make new alloys, such as bronze. More recently, people learned to separate metals that were very rare in their pure form in nature, such as aluminium and titanium, or not present at all, such as molybdenum and vanadium. From the mid-20th century, production of many of these increased hugely: the amount of aluminium produced since then exceeds 500 million tons, for example – more than enough to coat the whole United States in standard kitchen aluminium foil. The amount of iron and steel produced is many times that.

Since the mid-20th century, there has been a huge diversity of synthetic crystalline inorganic compounds made in materials science laboratories worldwide. Although these are not formally counted as 'natural minerals', in reality they are mineral substances – made by humans. These include substances such as boron nitride (harder than diamond, so used as an abrasive), tungsten carbide (used to make the ball in a ballpoint pen), synthetic garnets for lasers, graphene and many others. Indeed, at the last count, more than 200,000 of such artificial 'minerals' had been synthesised – 40 times more than the number of natural minerals! Through human ingenuity, during the 20th century, Earth has seen an increase in mineral diversity without parallel on any known planet.

▲ **Natural and synthetic minerals**
The upper two minerals are naturally occurring ores: rosasite, a rare copper-zinc ore from Mexico (left), and the green copper ore malachite together with limonite, an orange-coloured iron hydroxide, (right) from Cumbria, UK. Below are two synthetic 'minerals' that are both extremely rare in nature but now common at Earth's surface, thanks to human manufacture: pure silicon (left) and pure aluminium (right).

CONCRETE: EARTH'S ABUNDANT NEW ROCK

Concrete is so much part of our lives that we scarcely notice it, yet it is one of the most striking examples of a human-made rock. The recipe is very simple: crush limestone, mudstone and a little gypsum together, and apply high heat in a kiln to make cement. Mix one part cement powder with water and four or five parts of some kind of bulk filler – usually sand or gravel. The resulting slurry can be poured or spread into any shape desired, and in a few hours it begins to set. The result is a cheap, tough and durable synthetic rock.

Concrete has a long history: the ancient Romans made and used it, for example, discovering that if volcanic ash is added to the mix, the concrete can set underwater. But its growth to mass levels of use is a more modern phenomenon. The present-day recipe for Portland cement was developed in the 19th century; its use slowly growing during the Industrial Revolution. By the beginning of the 20th century, 30 million tons of concrete a year was made worldwide. This mushroomed to more than 10 billion tons annually by the year 2000, and to more than 25 billion tons a year currently. Around the world, half a trillion tons of this rock has now been made. The vast bulk of this has been made since 1950, and more than half of this in the last two decades. Today, concrete is not just an element of architecture, it is part of geology as well.

The planet is impacted by concrete in other ways. It takes a lot of energy to make the cement, and heating releases carbon dioxide from the limestone as it converts to lime for the cement. Altogether, concrete is responsible, for some 7 per cent of carbon dioxide emissions into the atmosphere, contributing to global warming.

Nevertheless, concrete as a 'new rock' is part of our living environment, with its own interest. The smoother types are in effect muddy sandstones (concrete skyscrapers have been called 'modern sandcastles'), while the coarser varieties are human-made conglomerates, their surfaces beautifully showing the pebbles within them. These are usually made of hard rocks such as milky white-vein quartz, flint in various colours, quartzite and many others. They tell their own ancient geological stories. A concrete wall or pavement can be a very good place to observe these rocks, and to try and guess their stories.

▶ **Roman concrete**
The Romans developed a form of concrete that was used in the construction of part of the Colosseum in Rome, Italy. The amounts used then, however, were trivial by comparison with today's use.

▶ **Concrete world**
Since the mid-20th century human use of concrete has increased more than thirty-fold, dominating the construction of our cities.

SAND: SMALL GRAINS, BIG BUSINESS

When looking at a concrete building, you are really looking at a sandcastle: sand makes up the bulk of concrete; the other ingredients (limestone and mudstone) used hold it together as a solid rock. To make enormous amounts (see pages 160–161), therefore, huge amounts of sand are needed, either exclusively or mixed with gravel.

That sand can be loose surface sand, or it can come from buried strata of more ancient sand, laid down millions of years ago. Not all sand is good for concrete production. The most obvious source – that in today's deserts – does not make good concrete; the sand grains have been so well rounded and smoothed by myriad impacts with other sand grains, that they do not bind well with other ingredients. The rougher, more angular grains of 'sharp' sand are needed, such as that found in river sands, or sands that have been carried by rivers, in river valleys or which have been conveyed into lakes or the sea.

The strata of ancient rivers are often good sources of sand and gravel deposits. Often, these are perched as flat-topped terraces above the modern river plain, representing former courses of the river at higher levels, before the river cut down to its lower, present-day level. These ancient river sands are often excellent resources, not least because they are often above the water table and do not become waterlogged when quarried.

▼ **A miscellany of sand**
Sand grains can vary greatly, as this arrangement shows. They range from smoothly rounded to sharp-edged, from spherical to elongated or irregular, and can be made of many different types of minerals. All of these factors affect the kind of sand that might be used for concrete-making and for other purposes.

RIVER HISTORY

Above the present-day river floodplain, there may be one or more river terraces –
higher, abandoned remnants of older floodplains of that river.

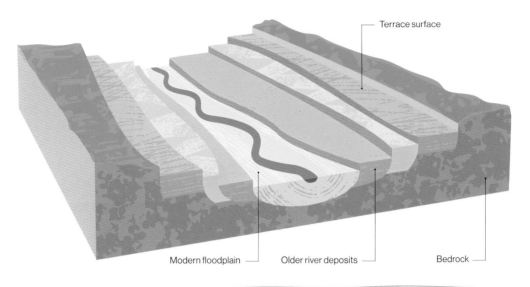

Terrace surface

Modern floodplain — Older river deposits — Bedrock —

Other ancient river sand strata date back to the ice age. Then, meltwater
streams spread extensive outwash plains from the ice as it melted in
summer. The glacial sand and gravel deposits laid down from the
meltwater torrents of those times now also contribute hugely to the
making of concrete.

The search for sand to help make concrete for the world's rapidly
growing cities goes further – to beaches and to the bottom of shallow
seas, from which the sand can be dredged or sucked up. Like any other
valuable resource where demand exceeds supply, sand can be supplied
from illegal or legal sources, and 'sand rustling' – often accompanied by
corruption – is a growing problem worldwide. It is difficult to comprehend
that such a simple substance is the basis of such big business that it involves
the criminal underworld – but sand is the main ingredient in building our
modern cities, and underpins our lives today.

One way to relieve the pressure on natural sand is to look for
other ingredients. Fly ash from power stations is one such substance.
Instead of being dumped in landfill sites, it can be mixed into concrete.
Concrete is increasingly recycled, too – crushed and added into new
concrete, which reduces its environmental footprint.

FIRING THE IMAGINATION:
THE SCIENCE OF BRICKS

Although bricks have now been overtaken by concrete as a building material, for thousands of years they were the world's most widely used human-made rock. And they are still around us in abundance – more than a trillion are manufactured globally each year. Originally, a brick was simply a shaped mass of mud, allowed to dry in the sun until it became hard. Such bricks were used in antiquity, especially in dry climates – some sun-dried bricks still stand today.

It was then discovered that fire-heating bricks made them even more durable. From about 5,000 years ago, bricks began to be produced in kilns; this new kind of rock contributed to the building of the earliest cities. The same kind of process was also used for making ceramics, a related type of synthetic rock.

    The recipe for today's bricks has been refined, but the basic formula remains the same. The bulk of the material is mud or mudstone, to which sand must be added (so the brick won't shrink too much when

▼ **A long tradition in brick-making**

The bricks here are made out of mud, drying and hardening in the hot sun. Such adobe bricks, if not as strong as modern fired brick, have nevertheless been used for thousands of years, and they continue to be made.

fired), together with a little calcium carbonate (limestone). Adding fossil carbon to the mix, as with black shales, is even better: this carbon burns as a built-in fuel during firing, saving on energy costs.

The firing of bricks (and ceramics) mimics the kind of process that takes place when magma comes into contact with rock strata underground. Geologically, it is a contact metamorphism of mudrock. But the human-made method happens very quickly compared with nature, taking just a couple of days rather than millennia; it also reaches temperatures 1,100°C (2,012°F) higher than is usually caused by magma contact. Indeed, brick-making temperatures can reach the melting point (helped by limestone content, which lowers the mix's melting point), so a small amount of synthetic magma is generated within the bricks during firing.

In the firing process, the minerals change from minerals typical of mudrocks (see pages 86–89) to entirely new ones. Most important in creating hardness is mullite, a mineral that combines alumina and silica, which forms from the transformation of clay minerals during firing. Its meshwork of long, thin crystals help hold the brick together. Mullite is rare in nature – named after the Scottish island of Mull, it is found in pieces of mudstone caught up in a prehistoric lava flow. The small amount of 'magma' in the brick – once cooled and frozen – also acts to bind the brick material, making it strong, durable and weather-resistant. In the classic red brick, the colour comes from the iron mineral haematite, produced during firing from other iron compounds in the mudrock. A very energy-demanding process, research continues into environmentally friendly ways of making bricks.

▲ **Durable brick architecture**
The mudbrick city of Rayen in Iran dates back a thousand years, and became a centre of trade and manufacture, and a defensive citadel. Many of its original structures still stand today.

▼ **A mineral helped by humans**
Mullite is a rare aluminium silicate, and it is a major component of cement and ceramics.

PREHISTORIC ORIGINS:
HOW HYDROCARBONS FORMED

For most of human history, the energy we used came from our own muscle power and from the burning of wood. Then, we began to exploit the muscle power of animals like oxen. As technology developed, windmills and waterwheels took energy from the wind and flowing water. These energy sources were endlessly sustainable, but had natural limits.

For the last few thousand years, some human communities also made use of a rock that burned – coal. Then, a few centuries ago, people learned to extract coal from the ground in larger amounts. In recent times, we've now developed methods of extracting oil and natural gas from rocks. The booming use of these hydrocarbons remains the central factor in powering our modern world. The energy it has supplied is prodigious – more in the past 70 years than from all energy sources in the preceding 10,000. Where does it come from?

Coal, oil and gas are all fossil fuels, derived from the remains of prehistoric plants and animals. In effect, they trapped the energy of fossil sunshine accumulated over hundreds of millions of years, energy that we are now releasing across just a few centuries.

▼ **Reconstruction of a coal forest**
The remains of many generations of trees in such an ancient forest build up and ultimately form a coal seam. The low-lying, waterlogged conditions help prevent decay of the wood, giving it a chance to be buried and preserved within strata.

Coal was mainly created from the remains of ancient forests and plant ecosystems such as peat bogs. Many forests live and die without leaving a trace, the remains decaying with carbon returning to the atmosphere as carbon dioxide. Coal forms where, after death, plants fall into waterlogged ground where decay is slow, so the remains are soon covered by other dead plant material. If the crust below slowly subsides, thick layers of such incompletely rotted vegetation pile up, eventually becoming tens or hundreds of metres thick. When these layers are deeply buried below more strata, the heat and pressure converts them into compressed coal seams, driving off natural gas in the process – which can be trapped in porous layers of rock above, now exploited by energy companies. The process continues today in the Florida Everglades, which are comparable to coal forests of the deep past.

Oil, by contrast, originates largely in the sea, starting as planktonic marine algae, the remains of which sink to the sea bed to form carbon-rich strata. Again, part of this dead plant matter needs be preserved within the sediment, so that it is then buried before it decomposes completely, its carbon dissolving back into the seawater. Such preservation is helped by stagnant, oxygen-starved conditions at the sea floor that hinder natural decay. Many of the world's large oilfields owe their origin to past warm climates when ocean circulation slowed and sea floor areas stagnated, allowing the buildup of thick deposits of organic-rich mud. These 'black shales', when buried, compressed and heated, release oil (a liquid, because of the fat-rich composition of the algae) and natural gas.

▲ **Coal seam**
The prehistoric forest that now comprises this coal seam was heated and compressed as a result of its burial deep underground.

BLACK CLOUD: THE CONSEQUENCES OF BURNING FOSSIL FUELS

Fossil fuels obtained from rock strata have driven the global economy for more than a century, and continue to be central to the way we live. They are energy-dense, and – particularly with oil and gas – easily transportable and convenient to use, powering industry, transport and homes, but not without consequences.

Burning fossil fuels releases carbon dioxide into the atmosphere, and with the large boom in human energy usage since the mid-20th century, the amounts have become massive – about a trillion tons. This is not all the carbon released (some has gone into the oceans, some into extra plant growth), and not all the extra carbon dioxide in the air has come from energy production: some has come from forest clearance, cement manufacture and other activities. But the bulk of it comes from burning coal, oil and gas.

How much is a trillion tons? In weight, that equals 150,000 copies of the Great Pyramid of Khufu in Egypt. In volume, it is a layer about a metre thick around the whole Earth, which is growing by 2 millimetres (0.08 inches) a month at current fossil fuel use. We know from measurements that it is almost 50 per cent more than the amount of carbon dioxide in the atmosphere before the Industrial Revolution began in the late 18th century. The fossil air buried in the compressed snowfall layers of Antarctica (see pages 152–153) show us that it is

▼ **Industrial exhaust from a furnace chimney**
Today, industrially expelled carbon dioxide in the atmosphere is equivalent to a layer of the pure gas about 1 metre (3.3 feet) thick around Earth, and it is growing by 1 millimetre every fortnight.

◀ Visible consequences of climate warming

As Earth's climate warms, the ice of the world's mountain glaciers and polar regions is melting at rates that now exceed a trillion tons of ice a year. This is causing the global sea level to rise, which leads to coastal inundation and erosion that reaches farther inland, endangering and destroying property.

almost 50 per cent more than has been in the air for at least 800,000 years. The last time there was so much airborne carbon dioxide was at least three million years ago, in the warmer climate of the Pliocene Epoch. What is this extra carbon dioxide in the air now doing?

Carbon dioxide is a greenhouse gas: that is, it traps infrared heat that would otherwise escape from Earth. This trapped heat has raised global temperatures by just over one degree Celsius in the last century. However, its effect on oceans is much greater, warming them by an estimated 14 zettajoules (a joule followed by 21 zeroes) each year. This energy capture is much greater than the amount of energy we obtained by burning fossil fuels in the first place (total energy use globally by humans per year is about half a zettajoule). One effect of this warming ocean is that the ice of Greenland and Antarctica has begun to melt: sea levels are rising by about 4 millimetres (0.16 inches) each year, and Earth's climate is changing profoundly.

INCREASING ACIDITY:
THE LIMESTONE ROCK CRISIS

The extra carbon dioxide from burning fossil fuels has effects that go beyond warming the climate. Ocean acidification is one effect, since part of that carbon dioxide ends up as carbonic acid dissolved in its waters. This has already changed ocean chemistry, tilting it towards more acid conditions: on average, surface ocean water has changed from a pH value of about 8.2 before the Industrial Revolution, to about 8.1 today. This might look like a small change, but the pH scale is logarithmic, so a change of one pH unit represents a ten-fold change in acidity. Consequently, oceans now are 25 per cent more acidic than before, and increasing

▼ Vulnerable organisms
Both corals (left) and pteropods (right) have skeletons made of aragonite, a mineral form of calcium carbonate that is harder than the better-known mineral calcite. It is also more soluble, and so vulnerable to the increasing levels of acidity of Earth's oceans.

This has significance for the formation of limestone rocks, as some of their most important building blocks – particularly animal skeletons, made of calcium carbonate – are affected by such acidic conditions. In life, animals such corals and pteropods ('sea butterflies') find it harder to make their skeletons, as they secrete a hard but more soluble calcium carbonate mineral called aragonite. Such skeletons are already becoming thinner and weaker, and so are both less bulky (so will form less limestone) and different in nature (weakened coral skeletons are more prone to damage by waves and animals that feed on them).

There is another effect taking place: increased dissolving of limestone sediment in the deep oceans. Over much of the ocean surface, there live microscopic plankton with calcium carbonate

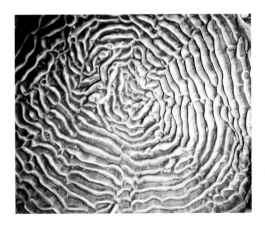

skeletons, such as coccolithophores (a single-celled alga) and foraminifera (amoeba-like protozoans). After they die, their skeletons sink in huge numbers towards the deep ocean floor. But in general, the deep ocean waters are more acid than those at the surface as more carbon dioxide is produced there by decaying organic matter. Once these tiny skeletons fall into these more corrosive waters, they quickly dissolve. Only in the shallower, less-acidic parts of the ocean – such as on the top of submerged volcanoes – can these skeletons survive to settle and form thick limestone oozes, above a kind of submarine 'snowline'. This snowline today, though, has already risen by hundreds of metres in parts of the oceans where human-produced carbon dioxide is penetrating, as plankton skeletons are increasingly dissolved. As these deep-sea oozes disappear, they will leave a legacy of decreasing limestone rock.

This kind of phenomenon has been seen in ancient strata, when natural episodes of increased carbon dioxide release and global warming took place during more volcanically active eras. Fifty-five million years ago, such an event took place, and has been tracked as a layer in which deep-sea limestones disappear, leaving only a residue of silica-rich muds that resisted dissolution. As today's global warming episode evolves, humans are creating our own 'limestone' gap as part of our legacy.

▲ **Multiple threats to corals today**

These corals are bleached a ghostly white because the water has become too hot for them, expelling the microscopic algae that normally live in their tissues and which give them their bright colours.

HYDROCARBON TRANSFORMATIONS:
AN EXPLOSION OF PLASTICS

▲ A sea of plastic
Plastic can collect in vast quantities, as seen here. Much plastic pollution, however, is invisible, including microplastics, such as the polyester fibres washed out of our clothes. These are now widespread across Earth, even on the deep ocean floor.

Among all of the new human-made substances, plastics – synthetic organic polymers – are rapidly becoming a ubiquitous component of sedimentary rocks forming today. Their manufacture and use started slowly in the early 20th century, with such materials as shellac and Bakelite. Then, with the invention of plastics such as nylon, polyethylene and polypropylene, their use really took off from the 1950s, after which two million tons were made each year. Production climbed steadily and is now approaching 400 million tons each year, meaning that we each generate our own bodyweight in plastic annually. Around 9 billion tons has been made, either in use or discarded (only a small part is recycled) – enough to envelope the planet in plastic wrap.

Plastics (numbering 20 main types) might be considered a new kind of synthetic 'mineral', manufactured using oil as a starting point. They are strong, light and decay-proof – qualities that make them highly useful. They are also cheap to manufacture, so are soon discarded after use – often after one use (e.g. soft drinks bottles) and often carelessly. Plastics litter the landscape, blown by wind and water into rivers. From there, millions of tons are carried into the sea each year, travelling far before becoming a component of new strata.

PLASTIC–ROCK HYBRIDS

Plastics have already formed new kinds of rock. One is called 'plastiglomerate', where melted plastic from beach bonfires has glued pebbles together. There is also 'pyroplastic': melted plastic globules that can look like real beach pebbles, except that they float. And plastic litter is already distinctively incorporated into naturally cemented 'beach rock'. Equally ubiquitous, plastics can travel invisibly to humans as tiny microfibres washed out of synthetic clothes textiles. Just one wash cycle can produce millions of such fibres, and these also find their way into rivers and oceans, where they can drift long distances before settling into deep-sea oozes. Today, any random handful of mud from the sea floor will likely have hundreds or thousands of microplastic particles within it. Plastics are a real signature of modern strata and, because they are so durable, will likely be fossilised as a permanent marker in future rocks.

New sedimentary plastics are of great concern as they interact with animals and plants, often in damaging ways. Birds and fish mistake plastic fragments for food, filling their stomachs with indigestible material. Coral colonies develop bacterial infections when covered in plastic trash. Even if all manufacture stopped tomorrow, the influx of plastics into the sea and sedimentary and biological cycles, will likely continue for millennia. This extremely serious situation requires careful study, so that it can be controlled.

Plastiglomerates
Origin:
Burning

Legacy:
Geological record
Burial
Chemical release
Biological effects

Pyroplastics
Origin:
Burning and weathering

Legacy:
Persistence in the environment
Chemical release

Plasticrusts
Origin:
Wave action

Legacy:
Persistence in the environment
Fragmentation
Plastic ingestion

Anthropoquinas
Origin:
Rubbish embedded in sedimentary rock

Legacy:
Geological record
Biological effects

DAMS AND RIVER RE-PLUMBING:
HUMAN INTERVENTION

Since Earth first evolved land, oceans and weather more than four billion years ago, rivers have played a key role in the making of rock strata, becoming the fundamental conduits transporting sediment derived from the erosion of land into the sea, where most rock strata form. Some of this sediment has been retained within the river system and fossilised in place, to form distinctive ancient river strata (see pages 138–139), such as can often be found associated with coal seams.

Over most of this time, rivers have behaved much in the same way, controlled by the physics of fluids and sediment particles. One revolution took place 2.5 billion years ago, when oxygen came into the atmosphere and changed the kind of minerals accumulating in rivers. A second took place 400 million years ago, when plants colonised the surrounding land. Plants' strong root systems, and the thick, cohesive soils they formed, constrained river water to flow through fewer, deeper channels migrating steadily across the floodplain, rather than frequently switching course.

Now that humans are transforming the landscape, rivers are being fundamentally altered once more in a third revolution, which undoubtedly will be reflected in the future rock strata left behind.

▼ **Dams hold back more than water**

The exterior view of a dam is of newly created expanses of water that can be used to generate power, for irrigation, or simply for pleasure. But beneath the water's surface large masses of sediment steadily accumulate, increasingly limiting the use and life span of the reservoir.

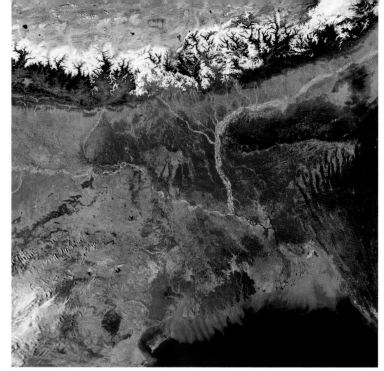

This satellite image
shows the extensive
delta of the Ganges-
Brahmaputra-Meghna
river system in India and
Bangladesh, upon which
more than 100 million
people live, and which is
threatened by rising sea
levels and subsidence.
The many dams built
on these rivers provide
water and power, but
starve the delta of the
sediment it needs to
replenish itself.

One change is the growing imprisonment of rivers within artificially
engineered concrete channels, especially in built-up areas. This results
in rivers effectively fixed in place, no longer able to migrate freely across
their floodplains to produce natural river strata.

Another change is the building of dams across most major rivers,
harnessing them to hold back water in reservoirs for urban supply,
hydropower, irrigation and sometimes for pleasure. However, in
relation to the rock-making process, dams also hold back billions
of tons of sediment that normally would be carried to the sea, that
instead piles up as thick layers of sediment behind the dams. This
amount of dam-trapped sediment is even greater as more sediment
now enters rivers due to increased land erosion associated with modern
deforestation, farming and urban construction.

Because dams act as huge sediment traps, much less sediment
reaches the sea, despite increased land erosion. This is crucial for major
deltas, usually sites of dense human settlement: deltas are now sinking
worldwide, threatening communities with inundation. A key factor is
that they are no longer nourished by sediment that could help them
withstand sea level rise.

Changes have occurred to the minerals that make up river sediments,
too, echoing the changes 2.5 billion years ago when a host of new
oxidised minerals appeared (and oxidisable minerals disappeared).
This time, the change is seen as human-made 'rocks' and 'minerals':
abundant fragments of glass, plastics, ceramic, bricks, concrete and
other materials, mixed in amongst the natural sediment.

SUBTERRANEAN: UNDERGROUND ROCK TRANSFORMATIONS

Characteristic features of rock strata are the traces of animal burrows, called 'bioturbation'. This kind of geological pattern first developed some 540 million years ago, when mobile, muscular animals first evolved in the evolutionary 'explosion' at the beginning of the Cambrian Period (see pages 132–133). Many types of fossil burrows exist, made by animals such as worms, crustaceans, sea urchins and others. Some are only a few centimetres in size. Larger ones can be a few metres deep, including the spectacular 'Devil's corkscrew' spiral burrows created by an extinct beaver 20 million years ago in what is now the US state of Nebraska.

These kinds of burrows, though, are dwarfed by the burrows now made by humans. We have entered – and permanently changed – the subterranean world of rocks far more than any other species in Earth's history.

Some underground journeys have been taken in search of resources such as coal, copper and gold in quarries and mines, which now extend up to nearly 5 kilometres (3 miles) underground (see pages 156–157). We have also extended our urban space for living and travelling

▼ **Unearthed marvel**
This extraordinary 'devil's corkscrew' is the fossilised spiral burrow of an early ancestor of the beaver. At around 3 metres (around 10 feet) long, it is among the deepest underground excavations made by non-human animals.

deep underground, as in the metro systems in major cities over the past 150 years. These are colossal and robust burrow systems: the largest (including Shanghai, London and New York) are hundreds of kilometres long, expanding out at regular intervals into underground stations, each with its own complex architecture. Underground space is now engineered for many reasons: for living, storage (underground salt caverns, and so on) and the disposal of hazardous waste.

We also penetrate deep underground without travelling there in person. Boreholes are drilled for oil, gas, water – or simply to discover what lurks deep underground. An example is the world's deepest borehole on Russia's Kola Peninsula, reaching 12 kilometres (7.5 miles) below the surface. The total length of boreholes drilled below Earth's surface is an estimated 50 million kilometres (30 million miles), roughly the distance from Earth to Mars.

Certain underground uses are more destructive, as with the nuclear bomb tests of the 1950s and 1960s, which left masses of shock-fractured, melted and radioactive rock – a human-made type of breccia. Other, more delicate types of new rock are associated with our underground activities: where underground water trickles through concrete tunnels, new types of stalactite and stalagmite form as 'calthemites', where mineral material from concrete dissolves and recrystallises as water drips.

These altered and new types of subterranean rock have a very good chance of being preserved for millions of years. Protected from erosion so deep below the effects of wind and rain, human-made rock traces may well be our most permanent legacy.

▲ **A nuclear crater**
This crater in the Nevada desert is the surface expression of one of the hundreds of underground nuclear test explosions carried out in the 1950s and 1960s. To a depth of up to 1 kilometre (0.6 miles), the rock beneath is shattered, partially melted and radioactive.

CITYSCAPES: URBAN ROCK STRATA

Towns and cities represent the newest and fastest-growing type of rock formations on Earth, where a huge variety of rocks may be viewed more easily than anywhere else. This is most obvious at surface level in buildings, with their constructions of concrete, brick, steel and glass, as well as in natural rock slabs (see pages 36–37), presenting endless geological variety.

▲ City rockscape
Modern cities are in effect intricate outcrops of rocks with natural components, such as sandstone and granite, and synthetic rocks that include concrete, brick and glass. Their composition is constantly evolving.

However, cities include other strata as well, sometimes hidden away from public gaze. The oldest are the rocky outcrops found in some urban areas, including those built on hilly terrain. Such exposures of ancient rock may be used to form distinctive and attractive urban features. The landscaping of Central Park in New York City beautifully shows rocks that reveal a long history, from being crumpled and metamorphosed 30 kilometres (18.6 miles) down in the roots of a mountain belt hundreds of millions of years ago, to being shaped by glaciers during the last ice age.

Cities may also be built on landscapes of 'soft rocks' deposited in the last ice age. Among the sediments themselves, hidden below streets and buildings, sometimes one can spot telltale signs of the landscape shape below the urban fabric, such as the flat surfaces of river terraces – ancient, abandoned floodplains – in cities such as London.

There are also remains of precursors to modern cities that now, as thick layers of rubble, underlie today's streets and buildings. Geologists call these rubble layers 'artificial ground': they are shown on geological maps, as are other kinds of sedimentary deposits. The older the city, the thicker the rubble layers will be. In cities with a long history, such as London and Rome, these layers can be tens of metres thick, representing generations of buildings, and are rich in archaeological remains. An assessment of rubble layers beneath London suggest that post-1950s rubble is now similar in mass to the rubble from Londinium going back to Roman occupation. Cities, as evolving masses of rock, are more dynamic now than ever.

Another rubble layer comprises a growing strata usually hidden from public view: landfill sites, typically locations such as disused pits and quarries, in which a city's waste is dumped. These can be enormous – much larger than the old middens of medieval town and cities. The new 'strata' contain a wide range of cast-off goods, from uneaten food to toys and electronic goods, usually with a high proportion of plastics. As they may contain toxic substances, modern landfill sites are often lined with and sealed by yet more plastic, so that nearby water supplies are not contaminated. This slows decay of the discarded objects, and will help fossilise them into the geological future.

▼ **Ancient foundation of a city**
The natural foundations of New York City, exposed as rocky crags in Central Park. Their complex history dates back over a billion years.

TECHNOFOSSILS: UNIQUE ROCKS

Two kinds of fossils are found in strata. Body fossils are familiar things like shells, bones, teeth and stems, and parts of the bodies of animals and plants. And then there are the traces left by organisms such as footprints and burrows. These are trace fossils, and it is among these one can sometimes find fossilised remains of constructions made by animals, such as wasp and termite nests.

Humans have excelled at building large, durable structures such as houses, roads and skyscrapers and one might also consider these as a kind of trace fossil, to be preserved as part of future strata. Of course, we make all kinds of other objects now – computers, cars, televisions, mobile phones, ballpoint pens, aeroplanes and much else – because of the ever more powerful technology that we have created. All these artefacts are robust, hard-wearing, decay-proof, and so perfectly capable of being fossilised in sedimentary layers, as technofossils.

Technofossils have some extraordinary properties when compared with ancient fossils. For instance, there are now many more different kinds of these than there are biological species today. Biologists do not know precisely how many species are alive today, but estimates are

▼ **Strata with technofossils**
This recently formed (and now eroding) beach rock in Spain is full of technofossils, including old bricks and furnace waste; the upper layers include plastic objects that date from the late 20th century.

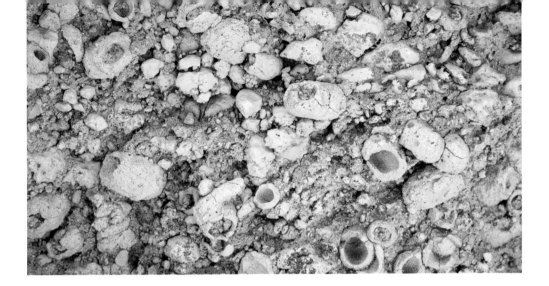

usually around ten million or so. Estimates of 'technospecies' numbers range into hundreds of millions of different kinds of human-made objects. We are living amid a remarkable proliferation of manufactured things, unique in Earth history.

Technofossils used to evolve very slowly, in their most ancient form. The flint implements used by our hunter-gatherer ancestors could stay the same over hundreds or even thousands of generations of the humans who made them. But as people settled, built towns and cities, and developed specialist trades, technological evolution quickened, as shown in the history of various implements unearthed by archaeologists. Now, new generations of computers and smartphones can appear every year. The evolution of technofossils has become completely detached from our own biological evolution, and is now taking place at scarcely believable speed. That, again, is unique in Earth history.

How might our technofossils be preserved, within strata into the far future? We can compare some examples with familiar fossils. Wooden objects such as chairs and tables will likely blacken and carbonise under millions of years of heat and pressure, to become chair- and table-shaped (though somewhat squashed) pieces of coal. Plastic objects too may become flattened and carbonised. Glass bottles might behave a little like natural volcanic glass, which slowly crystallises over time to become opaque. Iron objects will rust where there is water and oxygen, but if buried beyond the reach of oxygen, they will dissolve away in the groundwater, to leave technofossil-shaped holes in the rock. What might become of our really tiny and complex objects, though, like the circuitry of mobile phones? There is no close ancient analogue for that kind of thing, and this is among many such fascinating enigmas when considering what our future fossil record will look like.

▲ **Animal constructions as trace fossils**

These neatly constructed objects are nests built by solitary wasps, fossilised within ancient soil layers. Human-made constructions can be thought of as a similar, but much bigger and immensely more complex, phenomenon.

ROCKS ON OTHER PLANETS

SPACE ROCKS:
METEORITES

Before people learned to smelt iron ores, meteorites were almost the only source of iron on Earth's surface – a rare, prized material to mould into weapons and ornaments, such as the iron dagger wrapped with mummified king of Ancient Egypt, Tutankhamun. Past civilisations never dreamed that such curious rocks could fall from the sky: the realisation hit home, literally, to general astonishment, when meteorites fell on England and France in the late 18th century.

Meteorites are messengers from our most distant past that convey many tales of how the solar system was formed. They are mostly the debris left over from the construction of planets and moons, 4.5 billion years ago – though some rare examples have travelled more recently to Earth from the Moon and Mars.

The oldest meteorites can look more like a humble sandstone than an exotic extraterrestrial rock. Chondrites, made up of masses of chondrules (rounded mineral spheres) are thought to be the remains of packed melt droplets formed in dust clouds whirling around a just-lit Sun; some perhaps melted by violent shock waves and solar outbursts, some formed as splash droplets as embryonic planets grew and collided.

▼ **Meteorites from the dawn of our solar system**

Chondrite meteorites, with their distinctive masses of small rounded chondrules (solidified melt droplets) include mineral fragments from the dawn of the solar system, when the Sun was just beginning to shine.

▲ A Martian meteorite with fossilised life?

The Martian meteorite ALH84001 made headlines when possible fossilised bacteria were found within it. Alas, further study of these tanatalising micro-structures showed that they were more likely to be chemical, rather than biological, in origin.

◀ Patterns within iron meteorites

Iron meteorites, when cut, polished and etched, commonly show distinctive geometrical 'Widmanstätten' patterns that reflect the growth of the metal crystals.

Among the chondrules are tiny calcium- and aluminium-rich particles dating back 4.567 billion years. Other, even more minute particles have a chemistry so different that they seem even older – 'pre-solar grains' that have drifted in from other star systems. Other chondrites that formed a bit farther from the Sun include key ingredients that later formed the basis of life: carbon and water.

Alternative meteorite types include iron meteorites and stony meteorites, or achondrites. These are essentially fragments of growing planetary bodies smashed apart by collisions. The iron meteorites include some nickel, too, and represent the cores of these bodies as they grew, melted and separated into layers of varying density. They are the closest we can get to touching Earth's iron-nickel core. The achondrites, made of minerals such as pyroxene and olivine, are more akin to the less-dense material of Earth's mantle and crust.

A few achondrites have very particular chemical compositions that mark them as having come from the Moon or Mars, blasted off their surfaces by giant impacts, then drifting in space before chancing to land on Earth. One Martian meteorite, ALH84001, was famously proclaimed to contain traces of Martian life, such as fossil bacteria and organic chemical compounds. Alas! – these 'fossils' were later shown to have formed chemically, so the search for Martian life continues.

Most meteorites that land on Earth wear away and become unrecognisable. But meteorites can be hunted in places where they do not weather so quickly and are more easily seen. Wind-eroded surfaces of the Antarctic ice sheet and arid deserts are good hunting grounds.

TERRESTRIAL SCARS:
METEOR CRATERS ON EARTH

Compared to other solar system bodies, Earth shows few obvious meteoric craters. Around 190 have been discovered, some with meteorite impact rocks in and around them (see pages 146–147). Recognising them is not easy; most craters formed early in Earth's history, as the debris from the planet-forming process cleared up (by smashing into planets and moons). Since Earth is geologically highly active, most impact craters have long since eroded or been buried under sediment.

The largest known crater on Earth was at Vredefort, South Africa, and measured around 300 kilometres (186 miles) in diameter. The crater itself is long gone, but part of its foundations persist as huge rings of smashed and distorted rock that are visible from space. In the centre sit the remains of a 40-kilometre (25-mile) dome of rock, initially pushed deep into Earth on impact, only to rebound violently upwards.

Elsewhere, places where strata were forming are still preserved. Mineral grains with characteristic 'shock' features from the impact have been found in rocks hundreds of kilometres away; these are thought to be far-travelled ejecta from the Vredefort impact, with meteorite debris layers found as far away as Russia and Greenland.

The world's oldest-known meteorite crater is the Yarrabubba crater in western Australia. Like Vredefort, it is the eroded roots of a crater originally 70 kilometres (44 miles) across. Its age can be calculated because, among the smashed rocks of those crater foundations, there were tiny crystals of zircon, a uranium-bearing mineral used for dating rocks. These zircons had been there before the meteorite struck, but the impact remelted the crystal rims, resetting their internal 'atomic clock'. Careful analysis of the microscopic impact residues yielded an age of 2.3 billion years, plus or minus five million.

The dating showed a curious coincidence. Not only did it reveal the world's oldest discovered crater, it also indicated a time when one of Earth's earliest glaciations ended. Did the impact trigger that climate warming? The dust it raised may have darkened the ice-sheet surfaces, helping the Sun to warm and melt them; and if the meteorite struck an ice sheet, it could have dispersed huge amounts of water vapour – a greenhouse gas – into the atmosphere, causing warming. This is one of many tantalising hypotheses exploring how meteorites might affect the course of a planet's history.

ANATOMY OF AN ANCIENT METEORITE CRATER

A major meteorite impact, such as that which formed the Yarrabubba crater in Australia, may deform rocks of Earth's crust tens of kilometres below the surface, initially forcing them abruptly downwards, with major subsequent rebound. The resulting structure reflects both these processes.

Yarrabubba crater

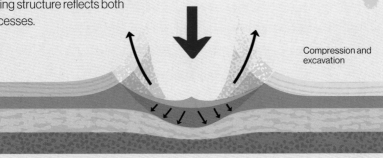

Compression and excavation

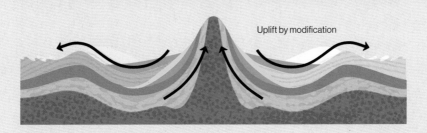

Uplift by modification

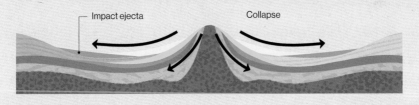

Impact ejecta

Collapse

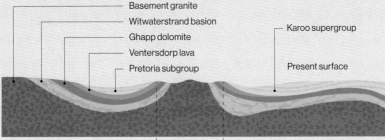

Basement granite

Witwaterstrand basion

Ghapp dolomite

Ventersdorp lava

Pretoria subgroup

Karoo supergroup

Present surface

Final crater

EXTRATERRESTRIAL IMPACT:
CRATERS ON OTHER PLANETS

Look up at the Moon with a pair of binoculars, and the powerful force meteorites play in shaping the rocky crust of a planetary body can be seen. The Moon's thousands of craters still exist because there are no plate tectonics, no water, no atmosphere and no weather to erode or bury them under sediment. Indeed, other than those covered by ancient lavas (see page 190), the greatest damage done to the Moon's craters is from other meteorites.

This continual pounding is now used as a means of telling the age of different parts of the Moon. Old areas of the surface, including old craters, have many newer craters on them, while the newest landscape has few or no craters. This crater-counting is used to estimate the age of the Moon's crater Clavius at four billion years old. The nearby crater Tycho, with its prominent rays, shows much less subsequent damage and is estimated at 100 million years old.

Crater-counting is used to help divine the history and age of the surfaces of the solar system's solid planets and moons. Mercury is similarly scarred: its craters resemble the Moon's. Venus has fewer craters, evenly scattered around the planet – a curious pattern hinting at a dramatic history (see pages 194–195).

▲ Craters of the Moon
Meteorite impact has been, and remains, the main geological process operating on the Moon. These impact scars range from enormous to microscopic, and many date back to the early days of the solar system.

The craters of Mars have been affected by the weather there: first, by running water, billions of years ago, when it was warmer and wetter. Some craters may have filled with water to become circular lakes, now long dried up. Now, Martian craters are subject to winds that blow in the thin atmosphere; small, active dune fields of Martian sand can be seen within them. Mars also might record a most gigantic impact, as its northern hemisphere is lower, flatter and younger than its more heavily cratered and mountainous southern hemisphere. These 'northern plains' are possibly the remains of a massive impact early in Mars's history, which would have almost torn the planet in two: a gigantic scar later forming a sea filled with sediment.

The gas giants, Jupiter and Saturn, do not retain traces of meteorite impacts on their fluid surfaces. But meteorites still strike them, attracted by their huge gravitational pull. In July, 1994, 21 fragments of the huge, broken-up Comet Shoemaker-Levy 9 slammed into Jupiter. Astronomers saw bright flashes as impact zones were instantly superheated to temperatures of 30,000°C (54,032°F), dark plumes of material from Jupiter's interior rising nearly 3,220 kilometres (2,000 miles). The plumes lasted for months before being dissipated by Jupiter's strong winds.

The moons of Jupiter, Saturn and dwarf planet Pluto all have patterns of crater damage. Some are ancient and heavily cratered, like Jupiter's moon Ganymede; some have fresher, lightly cratered surfaces, like Ganymede's neighbour Europa. These patterns are signposts to astonishingly varied histories in the solar system.

▼ **Mercury craters**

In these craters, the blue shades represent low-reflectivity material thrown out by the impact, which includes the graphite that is a distinctive part of Mercury's geology (see page 193).

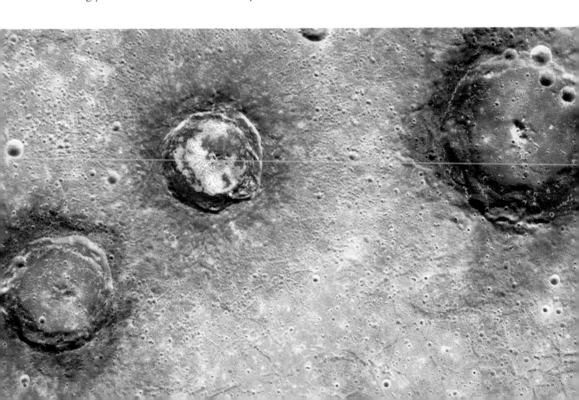

THE MOON'S ROCKS: ANCIENT HIGHLANDS AND BASALTIC SEAS

The fundamental patterns of the Moon's rocks can be seen by simply looking up at night – one can see lighter and darker patches, and these represent different rocks, with different histories.

The lighter areas are the ancient highlands of the Moon. They are the battered remains of its original crust, which formed four and a half billion years ago – the aftermath to the Moon's dramatic origins, as a mass of incandescent material flung out when the proto-Earth collided with a doomed Mars-sized planet, Theia. From this cataclysmic beginning, the Moon's surface was initially a deep magma ocean. As this magma ocean cooled, it began to crystallise: dense crystals of olivine and pyroxene sank down through the magma, while lighter crystals, mostly of a kind of feldspar called anorthite, floated up to form a thick layer of the pale igneous rock anorthosite, the Moon's first solid surface. Earth likely had a similar magma ocean and low-density crust after the impact too, though this was later completely refashioned by our planet's active tectonics, so none now exists. The Moon, much smaller, cooled and froze, so this

INSIDE THE MOON

The Moon, like Earth, has a layered interior, though it is almost (but not quite) geologically dead.

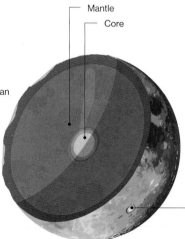

Mantle

Core

Moon covered by magma ocean 4.5 to 4.4 billion years ago

Moon origin by giant impact 4.5 billion years ago

3.9- to 3.2-million-year-old lavas now form the lunar maria

Many highland impact craters are more than 3.9 billion years old

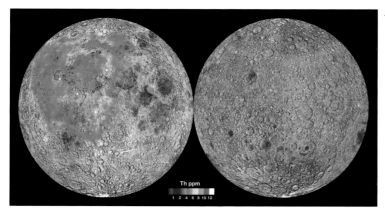

◀ **The Moon's hidden variety**

This map indicates how much thorium there is in the Moon's surface rocks. It shows the very irregular distribution of the most recently formed parts of the Moon's crust, with their distinctive 'KREEP' chemistry.

Th ppm
1 2 4 6 8 10 12

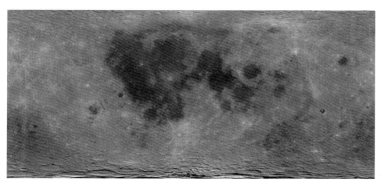

◀ **Face of the Moon**

The Moon shows bright, ancient cratered highlands, dark 'seas' of basalt lava, and some much younger meteorite craters – such as the Tycho crater in the south, with its conspicuous debris rays.

original crust still remains. Its surface was subsequently pulverised by meteorites, forming thick breccia layers of anorthosite fragments.
The Moon retained enough heat, though, to generate large amounts of magma and, over the next couple of billion years, this periodically flowed out, burying large areas of the lunar surface under basalt lavas. These form the Moon's dark patches, which were once called 'seas' and given names like the 'Sea of Tranquillity', but which are now known to be volcanic.

Lunar rock samples brought back from the Apollo space missions were found to contain high concentrations of potassium, rare earth elements and phosphorus (given the acronym KREEP). These elements did not fit into growing crystals very easily and so were left in the remaining magma, and when these last magmas crystallised, high amounts of those elements ended up in the rock. It was thought that this chemical composition should occur in rocks across the Moon. Yet when the chemistry of the Moon was mapped by an orbital satellite, these KREEP rocks were found to cover much of the near side of the Moon, including the landing sites, but only a small part of its far side. Why did these late magmas erupt in some areas, but not others? It is not yet known, and lunar scientists are still working to solve the puzzle.

MERCURY:
THE IRON PLANET

At first glance, Mercury resembles
Earth's Moon – dead, without air or
water, and heavily cratered. But its rocks
show specific features, the planet itself
having remarkable properties. More
than any other in the solar system, it
is metal-dominated. Its iron core, now
solidified, takes up three-quarters of its
diameter, so Mercury's rocky mantle
and crust are relatively thin. The planet
has been difficult to observe, so close to
the Sun's glare that it is hard to point
telescopes in its direction, while the
Sun's huge gravitational field makes
sending space rockets there difficult.
However, NASA's MESSENGER
spacecraft successfully reached Mercury
in 2011, and orbited the planet for
four years: much of our knowledge of
Mercury's surface rocks comes from
that voyage.

▶ **The faces of Mercury**
Mercury, so close to
the Sun's intense glare,
has been difficult to see
and image. It was only
mapped in detail once
NASA's MESSENGER
spacecraft managed to
get close to it in 2011. The
colours are light spectra
that relate to surface
chemical composition.

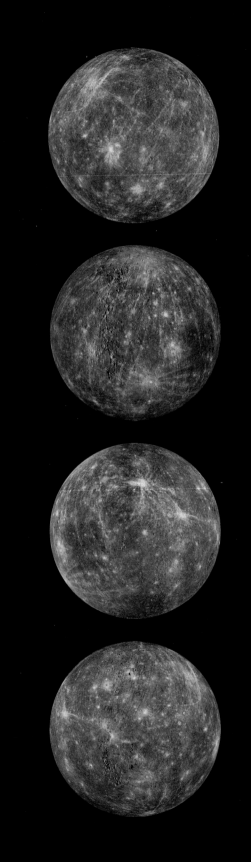

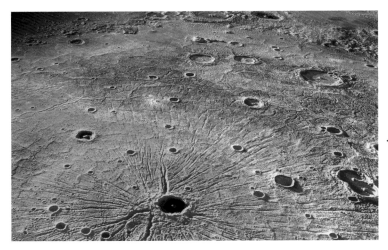

◄ **Mercury's Spider**
The mysterious 'Spider' on Mercury – more formally known as the Pantheon Fossae – is one of the planet's most striking and distinctive features.

Like the Moon, Mercury has a primordial crust, albeit one much more deeply buried beneath later volcanic rocks. It is only visible as scattered debris layers, thrown out long ago by huge meteorite impacts that excavated deep below the surface. Its composition, revealed by sensors on MESSENGER, was a major surprise. Unlike the pale anorthosite rock on the Moon, it is made of dark graphite, the form of carbon in pencil lead. Given Mercury's peculiar chemical composition, graphite was the material that floated to the surface of the planet's original magma ocean, while other growing crystals sank.

After this bizarre beginning, over the next billion years, Mercury became both highly volcanically active and bombarded by meteorites that etched its crater-scarred surface. Indeed, the greatest of those meteorite impacts likely helped the melting of the crust and mantle, to produce the voluminous amounts of magma that then flooded out as lavas across the planet. The volcanic rocks on Mercury's surface are mainly basalts, like those of the Moon and much of Earth, and also komatiites – very hot, fast-flowing lavas rich in magnesium. MESSENGER images also suggest that there had been explosive eruptions, which draped parts of the surface under volcanic ash layers.

Parts of Mercury's plains and craters are affected by tectonic folds that seemingly formed as the planet slowly cooled and contracted, to wrinkle the crust. The meteorite impacts may have also caused tectonic effects beyond the excavation of the craters themselves. On the other side of Mercury from the Caloris Basin impact, an area known as 'Weird Terrain' has a strangely distorted topography linked with this impact's shock waves. Within the Caloris Basin lies a distinctive pattern of radial fractures nicknamed 'the Spider' and more formally known as the 'Pantheon Fossae'. Near its centre is a small crater that may have formed from this later small impact.

▼ **MESSENGER to Mercury**
The MESSENGER (MErcury Surface, Space ENvironment, GEochemistry, and Ranging) spacecraft revolutionised our knowledge of Mercury.

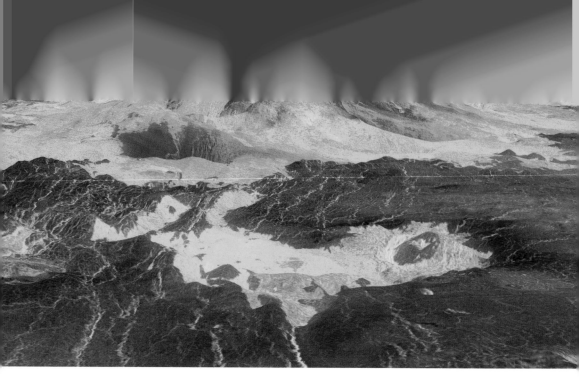

VENUS: A HIDDEN VOLCANIC LANDSCAPE

Venus has long been a mystery. With its surface hidden by thick cloud cover, it was imagined that luxuriant jungles might thrive below, harbouring all manner of exotic Venusian life forms. But when the first spacecraft penetrated the cloud cover in 1967, it found a lifeless inferno at 450°C (842°F). The searing temperatures maintained a crushing carbon dioxide atmosphere 90 times denser than Earth's, with clouds not of water, but sulphuric acid.

The rocky landscape beneath that hellish atmosphere is almost completely volcanic. No longer invisible, Venus's surface has been mapped by radar from orbiting satellites, revealing its extraordinary array of volcanic features. Some are wide lava plains and low, broad shield volcanoes like those on Earth, typically made of basalt rock, though on Venus they are lower and flatter. Other volcanoes are less Earth-like, such as clusters of small 'pancake volcanoes': low and flat-topped, they seemingly resulted from stiffer, more silica-rich lava. There are large circular structures called coronae, where plumes of

▲ **Hostile terrain**
Radar imaging from orbiting satellites has penetrated the thick, torrid sulphuric acid clouds to map the Venusian surface in fine detail, revealing spectacular volcanic landscapes such as this one.

rising material seemed to have pushed pieces of the crust upwards, breaking it along arrays of circular fractures where lava then oozes upwards. Lava channels – longer than those on Earth – extend for up to several hundred kilometres, perhaps because individual eruptions were on such a huge scale.

Venus's volcanic landscape bears only a few meteorite impact craters. This is partly due to the dense atmosphere, where smaller meteors burn up before reaching the ground. But there are not many larger craters, either, and these are seemingly scattered across the Venusian surface. It is clear, therefore, that this surface is relatively 'new' (only about half a billion years old), unlike the ancient surfaces of the Moon and Mercury. One hypothesis is that crustal renewal took place as a whole planet catastrophe. Venus shows no evidence of plate tectonics, so cannot get rid of its internal heat slowly and steadily, as happens on Earth. Rather, heat builds up beneath its single-piece crust, producing ever more underground magma, which eventually bursts out in a planet-wide conflagration, resurfacing the planet.

Although it has no plate tectonics, Venus has other tectonics. Where the crust has been stretched by movements of the mantle beneath, it has broken along parallel fractures to form arrays of rift valleys. Where it has been compressed, mountain chains have pushed up. The rocks have weathered as well, not as on Earth by water falling as rain, but by the hot acid and carbon dioxide-rich atmosphere, so different minerals – such as calcium sulphate and perhaps golden pyrite – form in Venusian soils, part of the bizarre rocky landscape of this strange and hostile world.

▲ **Venus hidden and revealed**

For centuries, Earth's astronomers saw only a blank, completely cloud-covered planet, even with high-powered telescopes (left). The radar images from satellite (right) revealed the landscape beneath, and revolutionised our understanding of the planet.

RED PLANET: THE ANCIENT VOLCANOES OF MARS

Mars is not so clearly dominated by volcanoes and volcanic rocks, as is, say, Venus. But its history has included much volcanism, most spectacularly in a huge, thickened region of crust called the Tharsis Bulge – the highest part of Mars, on which lie the largest and most long-lived volcanoes in the entire solar system. The continent-sized mass of igneous rocks that built up the Tharsis Bulge was likely created from magma derived from a huge long-lived plume of hot Martian mantle material. On Earth, tectonic plates move across such stationary plumes, so that chains of volcanoes (most famously, those of the Hawai'ian island chain) successively form over the 'hot spot', where the plume impinges on the moving plate. On Mars, this hot spot was always focused on the same place, and so much volcanic rock built up that their weight distorted the crust, causing fractures to extend far from the site of volcanism.

▼ **Mars's volcanic region**
A top-down view of the Tharsis Bulge on Mars shows a line of three large volcanoes and, to the side, the extraordinary volcanic edifice of Olympus Mons, the largest volcano in the solar system.

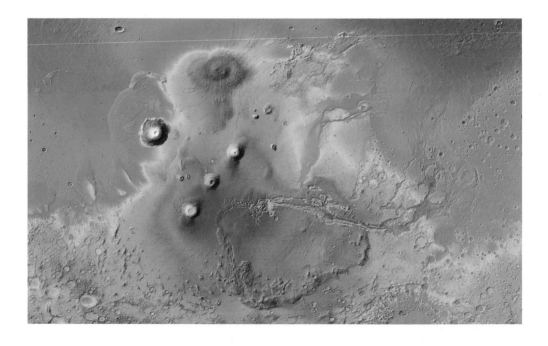

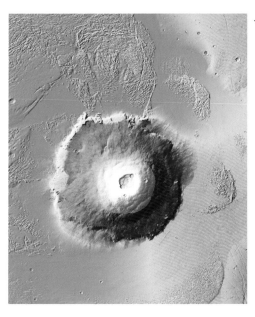

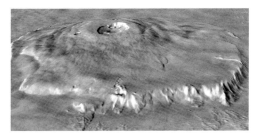

◀ **Olympus Mons from above**

Mars's relatively static tectonics likely caused the magma that formed Olympus Mons to follow a single route to the surface, contributing to its enormous size – it covers an area equivalent to that of Italy.

▼ **Olympus Mons from the side**

Viewed laterally, the high cliffs that surround this monstrous volcano are clearly visible.

Several huge volcanoes formed on the Tharsis Bulge. The biggest here (and in the solar system) is Olympos Mons. This shield volcano is more than 563 kilometres (350 miles) across and 26 kilometres (16 miles) high. Mauna Loa on Earth is also a shield volcano, but compared to Olympos Mons, it is terribly small, less than half the height – and only one-hundredth of the volume! Indeed, Olympos Mons is so big that, if you were standing anywhere on Mars, you could never see more than part of it. The best views are from space.

Olympos Mons began forming at least two billion years ago; the latest eruptions, measured by the extent to which they have been impacted by meteorites (see pages 188–189), are only a few million years old. Thus, it may not yet be extinct, and magma may be gathering for a future eruption. Like the many volcanoes on Earth, the Moon, Mercury and Venus, it is largely made of basaltic rocks. Being so huge and long-lived, it does not have a single crater at its top, but six overlapping each other, which collapsed following eruptions to form wide calderas. Curiously, and uniquely for Martian volcanoes, its base is marked by a cliff up to 8 kilometres (5 miles) high. How did this cliff form? Perhaps from landslides after the weight of the volcano caused its base to spread, the volcano originating at times when it was hemmed in by ice. It remains one of many puzzles.

While most observations of Mars's volcanoes come from satellite images, its volcanic rocks have been inspected close-up as well. When the Curiosity rover analysed minerals in the soils of Gale Crater, it found feldspar, pyroxene and olivine – all typical and familiar minerals of basaltic lavas that might be found on Earth. Basalt is clearly a fundamental planetary rock type.

MARS STRATA: TRACES OF A WARMER, WETTER PLANET

▲ Watery clues from Martian strata

Ancient sedimentary strata on Mars, now seen close up via the cameras of the Mars rovers, suggest that, billions of years ago, water flowed across the Martian surface.

Mars, like Venus, used to be thought of as a living planet, inhabited by intelligent beings. Nineteenth-century astronomers peered at fuzzy images of it through their telescopes, and some thought they glimpsed manufactured canal systems and seasonal changes in vegetation. Later, with better telescopes and spacecraft, it became clear that these were optical illusions.

Today, Mars is established as a frozen planet, with a very thin carbon dioxide atmosphere. Its water (and some of its carbon dioxide) is locked up as ice, most visibly as small polar ice caps. It is geologically active: winds blow in the thin atmosphere that carries sand grains that blast bare rock surfaces, or that pile up as still-moving sand dunes resembling dunes on Earth. Mars may be biologically dead, or it may have had some form of life, probably microbial – it may still have life, somewhere underground.

More than three billion years ago, Mars had a warmer, wetter past; this can be read from its rocks, both when observed from satellites and viewed by Mars rovers. But how could this have been, particularly as the Sun shone more faintly then? Most likely, its atmosphere would have been thicker, with more heat-trapping carbon dioxide, perhaps erupted from volcanoes that were then active. Whatever the case, clues to liquid water can be seen in its rocks. This early history may have included a sea that covered much of its current northern plains.

Mars strata also include fossilised river channels, a few with exquisitely preserved meanders. Some channels now form spectacular 'inverted' topography, where the effects of later wind erosion have made naturally cemented sandstone infills stand high above the rest of the softer weathering landscape. There are also valley networks carved into higher ground, which seem to have been cut by running water, and spectacular outflow channels, some larger than the Grand Canyon. Some of the water flow along these may have been catastrophic, as volcanism or meteorite impacts caused huge masses of ice to melt.

More details of water-lain strata are seen close-up by Mars landers and rovers. Some strata show traces of dunes swept along by running water, and there are also delicate layers of clay, laid down in ponds and lakes. Chemical structures are evident, too, such as the 'blueberries' – iron oxide concretions also associated with the action of water. So far, these seem to be traces of physical and chemical processes. Evidence of life, living or fossil, has yet to be found in these strata – but is avidly searched for.

▼ **River channels from an early Mars**
Satellite images show fragments of landscape that remain from more than three billion years ago, the result of Mars's feeble tectonic activity. These include well-preserved river channels – evidence of the former flow of water.

▼ **The imagined canals of Mars**
Giovanni Schiaparelli's 1877 map showing water-filled canals on Mars was beautifully drawn and finely detailed – but wrong. They amounted to a wishful over-interpretation of the fuzzy telescope images available in those days.

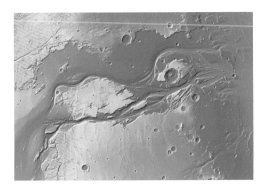

IO: THE MOST VOLCANICALLY ACTIVE BODY IN THE SOLAR SYSTEM

When first visited by spacecraft in 1979, Io – one of the Jupiter's moons – was seen to possess the most active volcanoes in the solar system. Amazingly, this was not altogether a surprise. Io is close to Jupiter, and some scientists predicted that it would be stretched and squeezed so much by the immense gravitational forces there, that these might melt its interior and cause volcanism. This prediction was vindicated, though even the scientists involved would have been surprised at how dramatic the volcanic consequences would be. That first flyby captured a mighty volcanic plume, rising 322 kilometres (200 miles) above the planet's surface. Volcanic eruptions proved frequent, but the volcanism, and the rocks it produced, emerged as more complex still.

The eruptions from some 400 active volcanoes, and the lavas and ashes they produced, have refashioned Io into a place of strange, multicoloured and extreme landscapes. The surface is very cold (–130°C, or –202°F), but the erupting magmas can emerge at very high temperatures, sometimes reaching 1,600°C (2,912°F). Originally thought to be of sulphur, they are mostly of silicate-based basaltic magma, like on Earth (some are sulphur-rich, though). The hottest magmas may well be of dense, high-temperature komatiites, as erupted on early Earth (see pages 126–127) and Mercury (see pages 192–193). There are almost no meteorite craters on Io, meaning that the surface is just a few million years old. Volcanoes keep reshaping it.

The volcanoes include paterae, large steep-sided depressions resembling caldera volcanoes on Earth, which form when the ground collapses beneath a volcano after the magma beneath is expelled. Io's volcanoes can be huge, often tens of kilometres across. The largest of

▼ **A multicolour moon**

The constantly renewed, volcano-dotted landscape of Io is a patchwork of different hues, the result of many overlapping spreads of volcanic ash and sulphur.

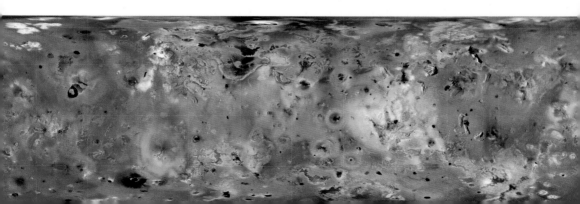

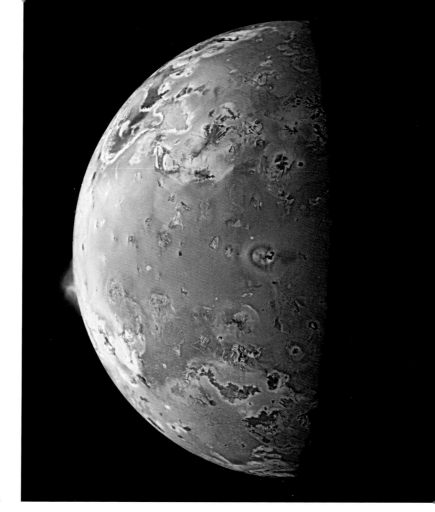

► An eruption captured on Io

This satellite image shows a major volcanic plume, hundreds of kilometres high, seen on the rim of this moon. At any one time, dozens of volcanoes are active on Io, with a diverse range of volcanic activity.

all, Loki Patera, is more than 200 kilometres (124 miles) across, filled with a lava lake that periodically forms a cooled lava crust, which cools and sinks, exposing the lava surface again. Lava lakes are rare on Earth, but common on Io.

The lavas can flow outwards in eruptions lasting months, to form lava fields extending from the volcanoes and fractures in the plains beyond. These seem to be like those on Earth – but larger and more fast-growing. Lavas can also erupt in brief, more dramatic episodes, as fire fountains and lava explosions, large and fierce enough to make Io appear brighter when seen from distant telescopes.

Very high volcanic plumes are actually less common types of eruption on Io. They mostly form from sulphur and sulphur dioxide deposits heated by magma, or released from erupting magma. These can rise to heights of 483 kilometres (300 miles), some carrying silicate magma particles up with them. The plume materials then slowly fall to Io's surface, blanketing it in multicoloured rings of sulphur and volcanic ash.

ICY EXTERIORS:
EUROPA AND CALLISTO

Around the orbit of Jupiter, and far from the Sun, the rock most common to Jupiter's 80 moons (other than volcanic Io) is water ice. One can imagine ice here as an igneous rock, crystallising from its 'magma' of liquid water, while these moons' surfaces may also have sedimentary ice grains, pebbles and boulders, and sheared metamorphic ice along tectonic zones. Such ice worlds are likely more abundant in the universe than the kind of silicate rock world that we live on, though they mostly do have silicate rock cores.

These ice worlds were once thought to be monotonous and largely alike, but we now know this is not at all the case. Two of Jupiter's ice moons, Europa and Callisto, demonstrate their astonishing diversity.

INNER STRUCTURE OF OUTER MOONS

Beneath the icy surfaces of Europa and Callisto there are likely hidden oceans and rocky cores.

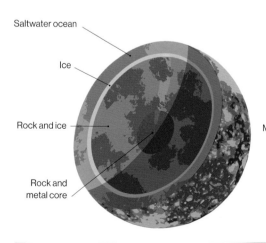

Saltwater ocean

Ice

Rock and ice

Rock and metal core

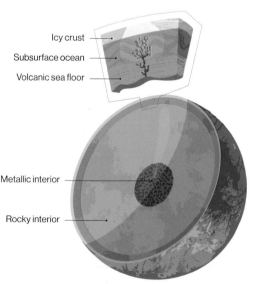

Icy crust

Subsurface ocean

Volcanic sea floor

Metallic interior

Rocky interior

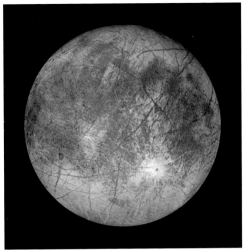

Europa, slightly smaller than our own Moon, is the smoothest known body in the solar system, with no large mountains. It also has very few meteorite craters, and so its surface is relatively 'new' – perhaps something like 100 million years old, thus dating back to the time of the dinosaurs on Earth. Its ice crust is nearly 100 kilometres (60 miles) thick, and overlies a water ocean of similar depth (so this underground Europa ocean holds more water than all of Earth's oceans). It is kept liquid by heating from the tidal forces exerted by Jupiter. And below that, the rest of Europa is made of rock, likely with a small iron core.

Europa's ice crust is not altogether smooth. Its surface shows a complex pattern of huge, moon-spanning fractures, highlighted by a deep red colour against the white of the ice in between them. These seem to represent some kind of 'ice tectonics', perhaps akin to Earth's plate tectonics, where the warmer, softer ice pushes upwards from below. There are also areas of 'chaotic terrain' – jumbled masses of ice blocks frozen together – that may represent where water, or warm ice, broke through from below. Plumes of water, instantly frozen to ice, have been observed fountaining above the surface – potential sites to seek the microbial life that might exist in that deep ocean. There may be regions, too, where the ice surface is shaped into steep spikes and ridges, as the ice evaporates from the feeble heat of the Sun.

Callisto is another icy moon, though with more ice and less rock than Europa – and in other respects it could not be more different. It has the most heavily cratered, and therefore probably the most ancient, surface of any body in the solar system. The craters overlap each other across its whole surface. Only the smaller ones have gradually been effaced by a kind of ice weathering, as the ice evaporates and recrystallises from the Sun's heat. Here there are no tectonics, no dynamic history, and little prospect of life.

▲ **Callisto and Europa**
(Left) Callisto's impact-battered but otherwise unaltered surface is a testimony to our Moon's lack of internal geological activity since its formation. (Right) Europa's geologically recent surface may lack mountain ranges and deep valleys, but it does not lack complexity, with the intricate meshworks of lines that traverse its ice crust.

READING THE LANDSCAPE: TITAN

The first astronaut to land on Saturn's largest moon Titan will experience a kind of geological dreamland, where everything is topsy-turvy compared to Earth. The bedrock is ice, as on the crust of Europa and Callisto, but there are some crucial differences:

- Titan's ice rises to form high mountains sculpted onto that ice crust, far below which – perhaps some 50 kilometres (31 miles) down – there is a frigid water-ammonia ocean.

- Titan, uniquely for a solar system moon, has clouds and rain, which are formed from hydrocarbons.

- The rain feeds rivers that carve canyons in the ice before flowing into lakes and seas of methane and ethane at about –180°C (–292°F).

Is this an oil geologist's dream? Well, these hydrocarbons cannot burn on Titan – not because it is so cold, but because there is no oxygen in the nitrogen-rich atmosphere. Just as on Earth, the hydrocarbon molecules can link up to form longer molecular chains, as you find in tar, so some of the Titan hydrocarbons are polymerised to form a thick haze in the atmosphere. Some of that haze clumps together to form sand-sized polymer particles, which are then blown by the Titan winds into spectacular dunefields. These will likely hinder our intrepid astronaut, as the particles are light and electrostatically charged – a little like trying to walk through a sand made of expanded polystyrene. There may be sand and pebbles of water ice, too, which should be kinder surfaces to traverse.

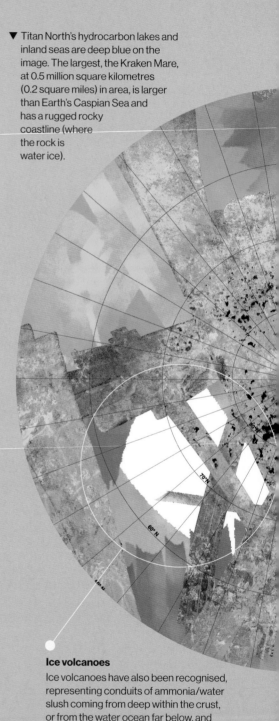

▼ Titan North's hydrocarbon lakes and inland seas are deep blue on the image. The largest, the Kraken Mare, at 0.5 million square kilometres (0.2 square miles) in area, is larger than Earth's Caspian Sea and has a rugged rocky coastline (where the rock is water ice).

Ice volcanoes
Ice volcanoes have also been recognised, representing conduits of ammonia/water slush coming from deep within the crust, or from the water ocean far below, and erupting and freezing at the surface.

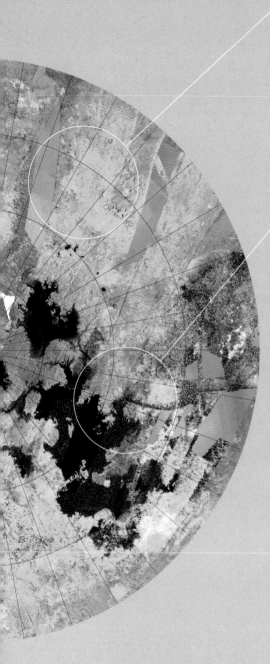

Titan rocks

So, what kind of rocks will be there, to be measured, hammered, collected? Well, one clue is that Titan is active. With mountains, valleys and lakes, and with relatively few meteorite craters on view, some kind of crustal tectonics must be continuously renewing that landscape. One can imagine that some of those mountains might include primary 'igneous' ice formed deep in the Titan crust, and then exhumed by erosion, while the depressions will accumulate sedimentary strata of polymers and ice grains. Those strata could well include ancient, petrified examples of the hydrocarbon dunefields now forming at the surfaces – just as on Earth there are both modern windblown dunes forming today and ancient fossilised ones in strata.

Rocky reservoirs

Titan's rock strata could be metamorphosed by heat (albeit the 'heat' here is still well below 0°C [32°F]) and pressure on being buried more deeply. Perhaps they are crumpled, too, into giant folds by whatever mysterious processes are raising the Titan mountains. The liquid hydrocarbons will not only be at the surface, but will percolate through porous, rocky reservoirs underground, just as oil and gas does on Earth – only on a much bigger scale.

▼ Artist's impression of dust storms

The dust, mostly tiny particles of solid hydrocarbon, is stirred up from the dune fields by storm winds at the planet's surface. Titan is, with Earth and Mars, one of only three bodies of the solar system known to exhibit such a phenomenon.

LESSER STATUS:
PLUTO

In 2006, Pluto had the indignity of being demoted from a planet to a dwarf planet. It is very small (much smaller than Earth's Moon; just one-fifth its mass) and very far away (at its most distant, 50 times farther from the Sun than Earth is), but its status is still debated. However, its own collection of rocks is unique, spectacularly visible when the spacecraft *New Horizons* flew past in 2015.

At about –230°C (–382°F), Pluto is a very cold and icy body. There are a variety of ices, with nitrogen ice (which covers most of its surface) as well as methane ice and water ice. The nitrogen ices form wide plains with a quilt-like pattern, such as the Sputnik Planitia, almost 1,000 kilometres (621 miles) across, this being the surface appearance of convection cells active in the soft, ductile ice, where it alternately sinks below the surface and wells up. These cells, each some 32 kilometres (20 miles) across, are constantly renewing the strange mobile 'rock' surface, which, from the lack of any meteorite craters, must be only a few million years old.

▼ **Pluto's nitrogen glaciers**
Even at only 40 degrees Celsius above absolute zero, Pluto's surface is dynamic, as shown with these slowly flowing nitrogen glaciers.

Rising above the nitrogen ice plains are steep and craggy mountains, up to several kilometres high, mostly made of water ice. These terrains are much older, the evidence being the many meteoritic impact scars, and represent the true 'bedrock' of Pluto. The mountains show signs of slow erosion, though, including at the edge of the plains, where kilometre-sized blocks seem to be dragged into the slowly flowing nitrogen ice, in trains of giant icebergs. These ancient areas are much darker than the plains, often with deep reddish-brown tones. The colours are probably due to complex organic chemicals formed by reactions in the exceedingly thin nitrogen-methane atmosphere, drifting down onto the rocky surfaces to give tar-like coatings.

Other areas, made of methane ice, have an extraordinary appearance and were dubbed 'dragonskin' or 'bladed' terrain when first discovered. The blades are regularly spaced ridges 0.5 kilometres (0.3 miles) high, formed as this kind of ice evaporates and re-condenses, reshaping the landscape into a spectacular form (such structures on Earth are called 'penitentes'). This terrain also has relatively few meteorite craters, so is one of the young landscapes of Pluto.

Remarkably, given how thin the nitrogen and methane atmosphere is now, there are small patches of windblown dunes as well – though these might have formed at times when the atmosphere was more dense.

Pluto's geology has proved to be unexpectedly various: it has been one of the many surprises the outer fringes of our solar system had in store for exploratory NASA missions.

▼ **An artist's impression**
Pluto's surface includes complex ridge patterns, some perhaps formed by crumpling of the ice, and some as kilometre-scale ice growths from the thin atmosphere.

ASTEROIDS: 'MINOR PLANETS' OF BIG SCIENTIFIC INTEREST

As well as the eight planets of this solar system, the six officially named dwarf planets (including former planet Pluto) and their combined total of 200-plus moons, millions more rocky bodies orbit the Sun. Most notable are the asteroids (or 'minor planets'). These are near enough to Earth for spacecraft to visit, while asteroids are occasional visitors to this planet, with sometimes catastrophic consequences. Most famously, 66 million years ago, a 10-kilometre (6.2-mile) asteroid impact killed off the dinosaurs and reset Earth's biology into the pattern still seen today. Scientists study asteroids keenly.

Most asteroids reside in the belt between Mars and Jupiter, in a zone where Jupiter's massive gravitational effects did not allow another planet to grow, the rocky debris of that zone instead colliding and breaking apart. The result is more than a million lumps of rock and metal larger than a kilometre across, and many more smaller ones (those less than a metre across are called meteoroids).

▼ Asteroid hunter
Japan's Hayabusa2 satellite, pictured here with the blue glow of its ion engine. In 2019, it sampled rock fragments and dust from the asteroid Ryugu, which were subsequently brought back to Earth.

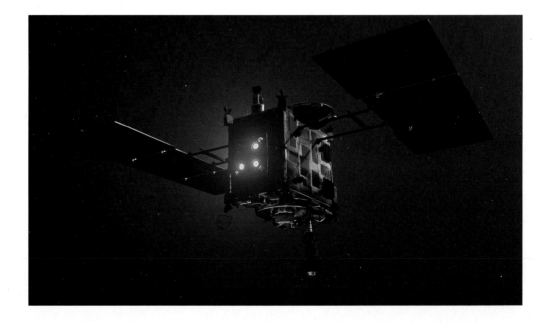

The largest ones, a few hundred kilometres across, are spherical, while smaller ones are irregular or simply form loose rubble piles. Many are impressively scarred by meteorite craters.

Asteroids are made primarily of similar kinds of rock to meteorites (that is, meteoroids that have landed on Earth: see pages 146–147). These consist of ancient chondrites, stony asteroids and iron-rich metal asteroids: some are broken-up parts of planets with a core and mantle/ crust that began formation before being smashed apart.

In 2010, the Hayabusa spacecraft made contact with the small asteroid 25413 Itokawa, which orbits relatively close to Earth. It was a mission in which all manner of misadventures happened: its solar cells were damaged by a flare from the Sun, making its ion engine malfunction; its lander missed the asteroid altogether; and the craft broke up on coming back to Earth – but by that time it had landed on the asteroid and managed to bring back a few tiny, precious fragments, which were recovered from the wreckage. With minerals such as olivine and pyroxene, these were shown to be like those of a chondrite meteorite. Amazingly, some – even though tinier than the width of a human hair – had minute micro-craters on them from space dust impacts.

More dramatically still, in 2019, the Hayabusa2 satellite 'bombed' another asteroid, Ryugu, firing a heavy copper impactor at its rubble-strewn surface to excavate a crater. Fresh material from this crater was subsequently delivered back to Earth for analysis, while the spacecraft itself is being directed towards other asteroids.

The interest in asteroids goes beyond science: as Earth's supplies dwindle, asteroids are being considered as possible sources of minerals and metals for future mining. This is a fascinating prospect, but our own planet has been spectacularly effective at producing a huge diversity of ores and minerals. It is difficult to imagine asteroids rivalling that!

▲ Planned encounter

An artist's impression of the European Space Agency's AIM spacecraft and its MASCOT-2 lander on the asteroid Didymoon. The landing is scheduled to take place in October 2022.

COMETS:
VERY OCCASIONAL VISITORS

Comets, those rare visitors to our night sky, used to be seen as mysterious portents of doom and destruction. More consistent observations show them to be less sinister, but just as fascinating. They are as much part of the solar system as planets and moons, but with a very different and distinctive pattern of appearance and disappearance.

We now know comets to be large masses of snow and ice mixed with rock and dust – huge 'dirty snowballs' – which, like the asteroids, are debris left over from the formation of the planets. Comets, though, mainly formed far from the Sun, beyond the orbit of Neptune, in cold regions where water vapour (and gases such as methane and nitrogen) condensed as ice-rich bodies, some big enough to be 'dwarf planets' like Pluto, hundreds of kilometres across. Most of these ice bodies are smaller, tens of kilometres across, and so usually invisible to us. It is only those possessing very elliptical orbits, sometimes venturing close to the Sun, that undergo the striking changes that make them spectacularly visible.

▼ **Colourful comet**
Comet Lovejoy passed by in 2015, and will return in about 800 years. Its well-developed coma, or head, is some 644,000 kilometres (about 400,000 miles) in diameter, and its greenish colour comes from fluorescence of diatomic carbon.

◀ **Impacted comet**
Comet 9P/Tempel
1, some 6 kilometres
(3.7 miles) across, was
famously targeted by an
impactor from the NASA
Deep Impact mission of
2005, producing a bright
explosion and throwing
out clouds of debris.

Then, the Sun's radiation causes the ices to vaporise, stream out from the comet and glow. The head of the comet forms a spherical 'coma' of ionised gas and dust, which can be huge (some have a diameter similar to the Sun's), part of which is stretched out by the solar wind as the comet's tail, which can be tens of millions of kilometres long. This amazing apparition lasts as long as the comet is near the Sun. Once it recedes, the lights go out – until its next flyby, which may be in hundreds of years.

In recent years, comets have been visited by spacecraft, landed on and even crashed into. Some surprising things have been discovered about the icy rock that makes up the comet. For all the brightness of the coma and tail, the solid heart of a comet turns out to be exceedingly dark, its surface covered by a black, tar-like layer of complex organic compounds, concentrated at the surface as the ice and dust below progressively blow into space with each orbit pass of the Sun. In 2005, the Deep Impact probe spectacularly crashed a heavy impactor into the comet Tempel 1, creating a crater 150 metres (492 feet) across and releasing a thick dust cloud, to reveal a mixture of dust and ice below the surface layer. The subsurface 'rock' was a fine and powdery mixture of ice, with tiny grains of silicate mineral, and clay and carbonate particles. This blend of materials shows how much mixing took place as the primordial solar system took shape.

Icy, comet-like bodies are thought to extend far out into space, within an 'Oort Cloud' that likely extends out at least halfway to the nearest star. These are a link as to what rocks might be expected in other star systems.

INTERSTELLAR GEOLOGY:
THE ROCKS OF OTHER STAR SYSTEMS

We can look up to see thousands of stars in the night sky, and our telescopes reach beyond this to show many galaxies. How many? The Hubble was once focused for a million seconds on an 'empty' patch of space, and detected about 10,000 galaxies in this single patch. From this, an estimate was made of 100 billion galaxies in the cosmos – and each galaxy contains something like a hundred billion stars. How much scope is there, thus, for rocks around other stars, and what form might these rocks take?

The first detection of an 'exoplanet' occurred in 1992, and more than 4,000 have since been found. They are so far away that we cannot yet see them directly by telescope. Instead, they are mostly found indirectly, by the way that a planet passing in front of a star causes its light to slightly dim, or its gravitational effect causes the star to slightly wobble. A 'rogue planet' has even been found, unbound from a star in the depths of space, detected by the way it 'micro-lensed' the light of a star it passed in front of (there are likely to be many such unbound planets). From these patterns, the approximate mass of a planet can be estimated, and the shape of its orbit around the Sun, its surface temperature, and sometimes faint indications of its surface chemistry. By the time you read this book, you may know a lot more, as the James Webb Telescope puts exoplanets in its sights.

We now know that many if not most stars have planetary systems, and they are often very different from our own well-ordered solar system. Many exoplanets have wildly looping orbits, taking them alternately near to and far from their sun. Many of the solid exoplanets will likely have some kind of tectonics and volcanism (for they must release their inner heat, just as Venus and Earth do, in their very different ways), and so magma and igneous rocks, including volcanoes and volcanic rocks, should be a common pattern.

With such exoplanet tectonics must also come metamorphism, and also ore and mineral formation in the rocks. Water, now detected far out into space, is likely even more of a universal constant, and present as liquid whenever local conditions allow. Many of those 'super-Earths' are likely to have oceans, some perhaps even being completely submerged waterworlds, and so sedimentary strata will be forming. From the experience of our own cosmic neighbourhood, yet more oceans will lie beneath exoplanet (and exomoon) ice crusts – a phenomenon we have yet to explore properly in this solar system.

▶ **Origin of a planet**
In this image of gas and dust swirling around a newborn star, the kink in the bright ring is thought to represent the gravitational effect of an infant planet.

▶ **Deep and ancient space**
A view of deep space taken by the Hubble telescope, showing galaxies so distant that, because of the time their light has taken to reach Earth, they represent the early history of the cosmos.

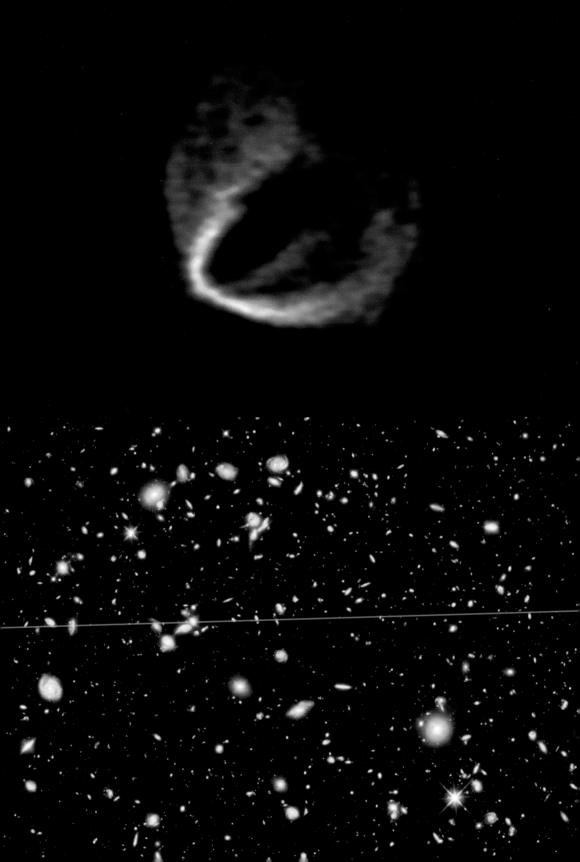

GLOSSARY

Ammonite A common spiral-shaped fossil found in strata of the Mesozoic Era.

Amphibole A dark iron-magnesium silicate mineral commonly found in basalts and gabbros.

Anorthosite A rock made mostly of the mineral anorthite, a kind of feldspar; it formed the primordial crust on the Moon and perhaps on Earth too.

Ash (volcanic) Particles of matter erupted from a volcano.

Asteroid A large rocky body in the solar system (though smaller than a planet).

Asthenosphere A 'weak' layer (due to a higher melt content) in the upper mantle, which enables plate tectonics.

Atoll A circular reef that grew around a subsiding island.

Banded iron formation Large deposits of iron ore that formed on the early Earth.

Basalt A dark, dense, fine-grained volcanic rock rich in iron and magnesium, which forms the ocean floors and is common elsewhere, including on other planets.

Beach rock Modern beach sediment that has been naturally cemented, for example by calcite, to form a hard rock.

Belemnite A cigar-shaped fossil common in strata of the Mesozoic Era.

Bioturbation Disturbance of surface sediment layers by crawling or burrowing animals, leaving a texture commonly preserved in rock strata.

Breccia A sedimentary rock made of coarse angular fragments of rock or mineral, naturally cemented together.

Calcite A calcium carbonate mineral common in limestones.

Caldera An area of subsidence that forms where a volcano collapses into an emptied magma chamber after a large eruption.

Chert A sedimentary rock made of fine-grained recrystallised silica; a common variety is flint.

Chondrite A kind of meteorite that formed early in the solar system's history, and which is made of tiny round chondrules (frozen melt droplets).

Clay The finest grade of sediment, commonly made largely of clay minerals – tiny flake-like minerals formed by weathering of other minerals.

Cleavage (mineral) The breaking of a mineral along regular planes that follow patterns of molecular structure.

Cleavage (rock) The splitting of a metamorphosed rock, such as slate, along planes formed by tectonic pressure.

Coal Layers of fossilised remains of land plants, now commonly used as a fossil fuel.

Cobble A roughly brick-sized sedimentary fragment.

Coccolith A kind of microscopic fossil that builds much of chalk.

Columnar jointing Regular fracture patterns that result from the cooling of a lava such as a basalt.

Comet An icy mass, usually far out in the solar system, that periodically becomes visible by flying close to the Sun.

Conglomerate A sedimentary rock made of pebbles and sand naturally cemented together.

Continent A mass of ancient, relatively silica-rich crust, the upper parts of which are usually mostly above sea level.

Coprolite The fossilised dropping of an animal, often rich in phosphate.

Coral A marine animal, often colonial, the skeletons of which are common as fossils; today, they form much of the framework of reefs.

Core The central part of Earth, mostly made of iron and nickel, molten in its outer part and solid at its centre.

Crater A circular depression excavated on a planetary surface: craters are formed by volcanoes or meteorite impact.

Crust The outer layer of Earth, thin in the oceans and thicker in the continents, and separated from the mantle by the Moho.

Crystal A form of mineral with a regular external shape that reflects the internal molecular structure.

Delta The mass of sediment dropped by a river as it enters a lake or sea.

Diatom A kind of microscopic fossil made of silica, common in some rocks.

Dune A large, regular, mobile structure formed by the interaction of sand with wind or water currents, forming distinctive strata common in ancient sandstones.

Dyke A near-vertical planar sheet of igneous rock, formed by the injection of magma along fractures.

Erosion The wearing away of rock by forces such as wind, rain and waves.

Erratic (glacial) A rock fragment that has been transported from its place of origin by a glacier or ice sheet.

Fault A rock fracture along which movement of rock masses has taken place, often generating earthquakes.

Feature mapping Working out underground rock structure from clues given by the topography.

Feldspar A silicate mineral with aluminium and sodium, calcium or potassium; overall it is the most common mineral at Earth's surface.

Flint A form of chert commonly found in chalk.

Fold A crumple in rock strata, typically from tectonic pressure.

Fossil The remains or impression of an animal or plant, preserved within strata.

Friable Crumbly and weak (which many rocks are).

Fulgurite A fused rock mass formed by a lightning strike on the ground.

Gabbro The coarsely crystalline equivalent of a basalt, where the magma cooled slowly underground.

Garnet A silicate mineral often found in metamorphic rocks, especially schists.

Geological Time Scale The time scale used by geologists for all rocks on Earth.

Geothermal heat Heat generated by rocks underground, for example by their natural radioactivity.

Glacier A moving mass of ice; on Earth it is made of water ice, but on other planetary bodies glaciers can have a different composition.

Gneiss A metamorphic rock greatly altered by heat and pressure.

Graben A slab of crust that has slid downwards between parallel faults.

Granite A coarsely crystalline igneous plutonic rock made largely of feldspar and quartz, commonly with some mica.

Greenstone belt Ancient terrains made of rocks of broadly basaltic composition.

Groundwater Water that is held between the grains of or within fractures in rocks underground.

Hornfels Mudrock that has been metamorphosed by nearby underground magma.

Hydrocarbon A compound of carbon and hydrogen – the basis for life, and also of fossil fuels such as coal, oil and gas.

Igneous rock Rock that has formed from the cooling and solidification of a magma.

Ignimbrite The deposit formed from a pyroclastic flow.

Intrusion An underground injection, typically of magma, that can cool to form plutonic igneous rocks.

Joint A rock fracture.

Karst A kind of terrain formed by dissolution of soluble rock such as limestone.

Kimberlite A kind of igneous rock that erupted from very deep within Earth; among its minerals are diamonds.

Komatiite A kind of hot, magnesium-rich lava, typically erupted on the early Earth.

Lava Magma that is erupted as a fluid onto Earth's surface, where it cools and solidifies.

Limestone A sedimentary rock made mostly of calcium carbonate (the mineral calcite) and commonly fossil-rich.

Lithosphere The upper part of Earth (the crust and uppermost mantle) that forms the moving plates of plate tectonics.

Magma Molten rock.

Mantle That part of Earth between the crust and the core; it is mostly solid.

Marble A metamorphosed limestone.

Metamorphic rock A rock (either sedimentary or igneous) that has been altered by heat and/or pressure.

Meteor A rock from outer space that burns up in the atmosphere.

Meteorite A rock from outer space that lands on Earth.

Mica An igneous or metamorphic silicate mineral with characteristic flake-like character caused by a well-developed mineral cleavage.

Mid-ocean ridge A raised area of the ocean floor where new ocean crust is continually formed as part of plate tectonics.

Migmatite A rock at the boundary between metamorphic and igneous (and so became partly molten).

Mineral An inorganic compound of more or less fixed composition, commonly crystalline, and the basis of rocks.

Moho The boundary between the crust and the mantle (in full, the Mohorovičić discontinuity).

Monolith (in geology) A single large rock.

Mud Fine-grained sediment that can combine clay and silt.

Mudrock A solid rock that was originally mud; also called mudstone and shale.

Ocean The part of Earth floored by oceanic crust, which is basaltic, relatively young and eventually recycled in plate tectonics.

Olivine A common rock-forming igneous mineral: an iron/magnesium silicate.

Ooids Spherical sand-sized grains of calcium carbonate, common in some limestones.

Ophiolite A fragment of ancient ocean floor pushed up onto land as a result of plate tectonics.

Orbicular texture A concentric texture found in some granites and gabbros.

Ore A concentration of useful minerals, especially metallic ones; ores frequently occur along mineral veins.

Pebble A sedimentary particle of rock or mineral between 4 and 64 millimetres (0.16–2.5 inches) in size.

Pegmatite A kind of very coarsely crystalline igneous rock, often with rare and diverse minerals; associated with granites.

Peperite A rock formed where magma steam-fragments the rock it is intruding into.

Peridotite A rock made mostly of the mineral olivine; common in the upper mantle.

Phyllite A mudrock that has been metamorphosed to a higher degree than slate; it is typically shiny because of tiny newly formed micas.

Plastiglomerate A new kind of rock formed of pebbles held together by melted plastic.

Plate tectonics The movement of pieces of the lithosphere relative to each other.

Plume A slowly rising column of mantle material: when it impinges on Earth's lithosphere, it can cause volcanism.

Plutonic rock A rock formed by magma solidifying underground.

Porphyry An igneous rock that has large, slowly grown crystals set in a finer and more rapidly cooled groundmass.

Precipitate (noun) A solid that forms from chemical compounds coming out of solution;

Pumice A light, very porous igneous rock formed by the rapid expansion of myriad gas bubbles during the eruption of viscous, silica-rich magma.

Pyrite An iron sulphide mineral known as 'fool's gold' because of its appearance.

Pyroclastic flow A hot, ground-hugging, turbulent flow of ash, rocks and gas from a volcanic eruption; the deposit left behind is an ignimbrite.

Pyroxene A dark-coloured iron-magnesium silicate mineral common in basalts.

Quartzite A metamorphosed sandstone.

Radiolarian A microscopic fossil with a silica skeleton, which can form sea floor sediments that harden to chert.

Radiometric dating Dating a rock or mineral by analysing the decay of radioactive elements within it.

Reef A biologically derived structure that can build up to form rock, usually a limestone; reefs today are dominated by corals.

Rhyolite A finely crystalline, rapidly cooled equivalent of granite.

Ripple Small-scale sedimentary structures in sands produced by air, water currents or water waves; they can be preserved in ancient rock strata.

Sandstone A rock made up of sand grains.

Schist A metamorphosed mudrock with abundant large mica crystals.

Scoria Lumps of chilled magma (usually basaltic) with a bubbly structure, though denser than pumice.

Sedimentary rock A rock made of sedimentary particles, such as a sandstone or limestone.

Seismic waves Energy waves that pass through rocks, commonly generated by earthquakes or explosions.

Shale *See* mudrock.

Silica Silicon dioxide, which in mineral form is quartz.

Silicate mineral The commonest rock-forming mineral; based on a silicon–oxygen chemical combination.

Sill A broadly horizontal injection of magma (often along stratal surfaces) that cools to form a sheet of plutonic igneous rock.

Silt Fine sediment, of grain size between sand and clay; it can form a siltstone, or be a component of mud and mudrock.

Sinkhole A depression in the landscape, often roughly circular, formed as a result of underground rock collapse.

Skarn A metamorphic rock formed by magma coming into contact with limestone.

Slate A mudstone metamorphosed at relatively low temperature and pressure, characterised by well-developed slatey (rock) cleavage.

Stalactite A rock made of precipitated calcium carbonate that hangs from a cave roof.

Stalagmite A rock made of precipitated calcium carbonate, growing up from a cave floor.

Stratum (plural strata) A layer in a sedimentary rock.

Stromatolite A layered limestone structure built by microbes, and among the oldest fossils on Earth.

Subduction zone Where ocean crust is being destroyed by sliding down into the mantle as part of plate tectonics.

Suevite A rock formed by meteorite impacts.

Technofossil An object made by humans capable of being fossilised into future strata.

Tectonics The movement of rocks within Earth that forms structures such as faults.

Thrust fault Where one block of rock slides over another along a low-angle fault plane.

Till (glacial) Also known as 'boulder clay', a mixture of mud, sand, pebbles and boulders smeared across the landscape by moving ice.

Topography The shape of the landscape; often controlled by the underlying rock structure.

Trilobite A marine animal fossil typical of the Palaeozoic Era.

Turbidite A rock stratum deposited from an underwater turbidity current.

Unconformity The discordant contact between older eroded rocks and younger overlying strata, which often marks a large geological time gap.

Vein (mineral) A rock fracture filled with minerals deposited from underground hot water; may include ores.

Weathering The breakdown (especially chemical) of rocks at the land surface.

Xenolith A detached lump of rock incorporated into a magma.

Zircon A mineral (zirconium silicate) formed in igneous rocks, and used by geologists for radiometric dating.

INDEX

PICTURE CREDITS

ACKNOWLEDGEMENTS

Heartfelt thanks to my many mentors and colleagues through the years, who have helped me explore the infinite worlds to be found within rocks